# Systems Biology and Bioinformatics

## A Computational Approach

# Systems Biology and Bioinformatics

## A Computational Approach

Kayvan Najarian
Siamak Najarian
Shahriar Gharibzadeh
Christopher N. Eichelberger

CRC Press
Taylor & Francis Group
Boca Raton London New York

CRC Press is an imprint of the
Taylor & Francis Group, an **informa** business

CRC Press
Taylor & Francis Group
6000 Broken Sound Parkway NW, Suite 300
Boca Raton, FL 33487-2742

First issued in paperback 2017

© 2009 by Taylor & Francis Group, LLC
CRC Press is an imprint of Taylor & Francis Group, an Informa business

No claim to original U.S. Government works

ISBN-13: 978-1-4200-4650-2 (hbk)
ISBN-13: 978-1-138-11803-4 (pbk)

**Visit the Taylor & Francis Web site at**
**http://www.taylorandfrancis.com**

**and the CRC Press Web site at**
**http://www.crcpress.com**

# Dedication

*To our families whose patience and encouragement
gave us all we needed to write this book.*

# Contents

# Preface

## WHAT IS SYSTEMS BIOLOGY?

The book you have in hand is written to cover the main computational methods used in two very vibrant and dynamic fields of science: bioinformatics and systems biology. Although the notion of bioinformatics might be the younger of the two fields, the ambiguity surrounding the exact definition of systems biology made us start the preface with a simple yet important question: "What is systems biology?" After we answer this question on systems biology, we will ask a similar question regarding bioinformatics. We start answering the above-mentioned question using a historical standpoint.

### HISTORY OF SYSTEMS BIOLOGY

Analyzing biological data has long been a dynamic field of science. The main objective of such analyses is to process the measurements captured from the biological system under study to describe the functions and behavior of the system. However, until two decades ago, almost all available measurements for any biological system were "macrolevel" recordings that were unable to register the physical/chemical properties of individual molecules. Consequently, the resulting analyses provided models of the biological systems that explain the function and behavior of the systems in macrolevel that were often insufficiently accurate or specific. Although at the time the science of forming these macrolevel models were sometimes referred to as "systems biology," as we will see soon, this name was later given to a new concept that was born in the 1990s.

Since the early 1990s, biological sciences witnessed a revolution in high-throughput molecular measurement systems. This revolution owed its impact to an unlikely marriage between molecular biology and semiconductor technology. The resulting technologies allowed hundreds and thousands of biological molecules in a single experiment. Collection of the same amount of information collected in a single high-throughput experiment would have taken months or even years using conventional methods. These high-throughput technologies, such as DNA microarray machines, produce thousands of data points in a single experiment, and since the use of these technologies is now widespread, a significant amount of molecular data is being produced everyday across the world.

The novel high-throughput experiments have provided us with such an abundance of biological data at molecular level that new computational methods are needed to process the data and produce answers for new questions in biology. Using these data, nowadays, many bold hypotheses and theories in biology can be tested provided that the data produced by the wet laboratory experiments are properly analyzed and modeled. Above all, instead of searching specific answers for very specific questions, biologists are more interested in "holistic" approaches in which the main attention is given to the interactions among all participating biological elements at the molecular level as opposed to the functions of particular elements as stand-alone units.

DEFINITION OF SYSTEMS BIOLOGY

With the detailed introduction given above, we are ready to form our own "informal" definition of systems biology. As it is clear from the history, there have been very many definitions provided for systems biology and the definition given here is by no means intended to cover all existing definitions. As referred to in this book, systems biology is defined as a set of computational methods that intend to integrate all molecular measurements captured from the elements of a biological system to create a model that describes and predicts the overall behavior of the system. This definition, although a bit long and "wordy," clearly explains the reasons and motivations for forming systems biology models.

## WHAT IS BIOINFORMATICS?

Bioinformatics, which shares some of its main concepts and definitions with systems biology, focuses mainly on some major problems in molecular biology including study of function and structure of proteins, identification of the functions of each gene, and extracting information from human genome as well as the genome of other organisms. Bioinformatics can be simply defined as a collection of computational methods used to discover, process, and analyze patterns in genomic and proteomic data sets. These methods typically use statistical and probabilistic models as well as models rooted in machine learning.

Computational method bioinformatics are rather more established compared to systems biology, and although the present book will cover both topics in detail, due to the profound impacts of systems biology on the future of medicine and drug design, the main emphasis of the book will be on the computational methods in systems biology.

## PROMISES, CHALLENGES, AND FUTURE

The impact of both bioinformatics and systems biology on drug design and medicine is tremendous. Consider a simple yet extremely important issue encountered in drug design: late ineffectiveness. It is known that many drugs can show promising results for a certain period of time, but after a while, the disease reemerges. A few such systems have been closely observed and carefully analyzed. The results indicate that these biological systems are flooded with feedback and feedforward loops, and the direct interactions between the drug and molecules involved in the disease are sometimes not more important than the indirection interactions through the loops. Technically speaking, as a result of these loops, before the system shows its steady-state response to the drug, it can exhibit dramatically different transient patterns. Such observations further emphasize the importance of systems biology models that can produce a holistic picture of the system that considers both direct interactions and indirect (through loops) interactions.

As another example, let us consider another major challenge in drug design: choosing the right drug for a new strain of a viral disease. Since retroviruses lack a stable genetic blueprint, they frequently mutate inside the host cells. Although many such mutations do not alter the structure of the main proteins involved in the life cycle of the virus significantly, some of these mutations might change the structure of some of these proteins. Now, if the drugs patients are using were designed to target, bind, and inhibit these important proteins, after these structural mutations, the mutated proteins may no longer bind with the drugs. This means that the existing drugs are no longer useful, and a search for new drugs has to be initiated. There are a number of bioinformatics techniques that search for the drugs that best bind with a given binding site on a protein.

## ABOUT THE BOOK

This book is intended to be used as the textbook for either senior undergraduate level or first year graduate level courses. The main background needed to understand and use the book is college level calculus and some familiarity with probability and statistics. Although some knowledge of linear algebra would also be helpful in understanding the concepts, linear algebra is not a prerequisite. The book attempts to describe all mathematical concepts and techniques without getting stuck with the details.

The book does not assume that the reader is an expert in biology either and starts with a brief review of cell and molecular biology. The book also briefly describes some of the main high-throughput measurement technologies and their applications in biology. However, the readers are encouraged not to limit their biology readings to this book. The use of a textbook in cell biology as a reference book would help the reader better understand some of the advanced biology concepts covered in this book.

Each chapter also contains some exercises in sections called "Problems" that give readers the chance to practice the introduced computational techniques. Some of the problems are simply designed to help students practice and improve their understanding of the computational methods, while the rest are real-world practical problems using real data from biomedical systems.

## OVERVIEW OF CHAPTERS

Chapter 1 gives a very brief review of cell biology. This chapter also briefly reviews the main ideas of molecular biology. The concepts in Chapter 1 are written in a language understandable by readers with no advanced background in biology.

Chapter 2 describes the main high-throughput bioassays used in systems biology. The mechanism and applications of these measurement systems, such as DNA microarray and PCR, are described in detail. The technical descriptions are kept at a level that is useful for both engineers and biologists.

A review of the computational methods and mathematical theories that are commonly used in systems biology and bioinformatics is given in Chapter 3. These methods and theories include random variables and probability theory, Bayesian theory, test of hypothesis, expectation maximization method, maximum likelihood method, and system identification theory.

Once we are familiar with cell biology, measurement systems, and some basic computational methods, we are ready to apply some computational techniques in bioinformatics to analyze the resulting data. In Chapter 4, the focus is given to computational structural biology. Chapter 4 is the first of three chapters dedicated to principles of bioinformatics. This chapter is dedicated to some of the oldest problems in bioinformatics including protein and RNA structure prediction and protein folding and dynamics.

Chapter 5 covers the computational methods that analyze the molecular sequences, such as the sequences of amino acids in DNA, and extracts useful patterns from these sequences. This chapter also discusses some of the major databases and software tools for sequence comparison and matching such as BLAST and FASTA.

Chapter 6, which is titled "Genomics and Proteomics," essentially covers all other bioinformatics methods applied to process genomic and proteomic databases. These include the methods to identify (1) a gene in DNA sequence, (2) a promoter region in DNA sequence, (3) a motif in protein sequence, and so on. Chapter 6 is the last of the three chapters focusing on the mainstream bioinformatics topics, and starting from Chapter 7, the focus will shift toward systems biology.

Discovery of differentially expressed genes across DNA microarray samples taken at different experimental or clinical conditions is an extremely important and commonly encountered application in today's biology and medicine. Chapter 7 is dedicated to methods that identify differentially expressed genes or proteins across two groups of samples. This chapter starts with description of simple methods such as $t$-test and moves to advanced methods such as the mixture model method.

Chapter 8 is the first chapter in which we form holistic models of all elements (genes or proteins) involved in a system and identify the interactions among the elements in order to describe the overall functions of the system. This chapter describes Bayesian models that estimates regulatory networks of genes or proteins that alter and adjust the expression level of a certain gene and/or protein. The models discussed in this chapter are all static, i.e., all interactions are steady-state, and no transient interactions are taken into consideration.

Another family of static models that are useful in making systems biology models is created using metabolic control theory. This theory and its application in systems biology is described in Chapter 9.

While static models can be very helpful for certain applications, in some applications, the transient behavior can be as important as the steady-state behavior if not more important. In such applications, the models described in Chapters 8 and 9 cannot provide the solution. Chapter 10 describes the principles of system identification theory and explains how this theory must be combined with methods such as ICA and factor analysis to create more reliable dynamic gene/protein regulatory networks.

Chapter 11 describes an important topic in molecular biology that has been playing an increasingly important role in identification of biological pathways and regulatory networks. Gene silencing technology has provided means to slice sophisticated networks into simpler ones and study these less complex networks of genes to explore the relationships among the elements. In Chapter 11, we explore mathematical formulations that allow systematic formation and analysis of biological networks using what RNAi and gene silencing technology has provided us with.

In Chapter 12, we rather change our standpoint at systems biology and bioinformatics as "applications" for existing computational methods and treat them as "inspirations" for novel computational paradigms. Chapter 12 discusses the computational techniques that are formed as in silico emulation of biological systems, in particular, in the molecular level.

Chapter 13 introduces some of the main software tools and databases commonly used in bioinformatics and systems biology, including GEO, GenMAPP, MATLAB® Bioinformatics Toolbox, and so on.

The final chapter of the book, Chapter 14, briefly discusses some of the future trends and directions of bioinformatics and systems biology. Some of these trends, which may not be achievable in the immediate near future, will have substantial impact on the way medicine and drug design are defined.

MATLAB is a trademark of the MathWorks, Inc. and is used with permission. The MathWorks does not warrant the accuracy of the text or exercises in this book. This book's use or discussion of MATLAB software or related products does not constitute endorsement or sponsorship by The MathWorks of a particular pedagogical approach or particular use of the MATLAB software. For product information, please contact:

The MathWorks, Inc.
3 Apple Hill Drive
Natick, MA 01760–2098 USA
Tel: 508–647–7000
Fax: 508–647–7001
E-mail: info@mathworks.com
Web: www.mathworks.com

# Acknowledgment

The authors thank Soo-Yeon Ji and Rebecca Smith for providing the authors with their invaluable feedback on some chapters of this book. The detailed feedback from these individuals helped the authors improve the chapters of this book. In addition, the authors thank Dr. Alireza Darvish for his contributions to Chapters 9 and 10.

Moreover, the authors would like to thank all hospitals, clinics, industrial units, and individuals who shared their biomedical and nonbiomedical data with the authors. In each chapter, the sources of all contributed data are mentioned, and the contributions of the people or agencies that provided the data are acknowledged throughout the book.

# Authors

**Kayvan Najarian** is an associate professor in the Computer Science Department at Virginia Commonwealth University, where he works on biomedical signal and image processing and biomedical informatics. Previously, he was assistant professor at University of North Carolina at Charlotte. Dr. Najarian has a Ph.D. in electrical and computer engineering, which he completed with an award-winning thesis. He has taught extensively since 2000 at both the undergraduate and graduate levels at University of North Carolina at Charlotte and Virginia Commonwealth University, including the creation of new courses and development of the curriculum. Dr. Najarian has been published six times in the *Wiley Encyclopedia of Biomedical Engineering*. He coauthored *Biomedical Signal and Image Processing*. He has published more than 100 journal articles and conference papers. He is a senior member of the Institute of Electrical and Electronic Engineers, is a reviewer, referee, or member of dozens more committees, journals, and councils.

**Christopher Eichelberger** is the director of the Software Solutions Lab within the College of Computing and Informatics at the University of North Carolina at Charlotte. His research work has been centered on the intersection of life sciences and computer science, with publications in artificial life, artificial intelligence, complex adaptive systems, data mining, and simulated protein computing.

**Siamak Najarian** serves as the full professor and dean of the Faculty of Biomedical Engineering at Amirkabir University of Technology. He has completed his Ph.D. in biomedical engineering at Oxford University, UK, and had a postdoctoral position at the same university for a year. His research interests are the applications of artificial tactile sensing (especially in robotic surgery) and design of artificial organs. He is the author and translator of 22 books in the field of biomedical engineering, 7 of which are written in English. Dr. Siamak Najarian has published more than 120 international journal and conference papers in the field of biomedical engineering.

**Shahriar Gharibzadeh** received his M.D. degree in 1992 and his Ph.D. degree in medical physiology in 1998, both from Tehran University of Medical Sciences. He has been with the Faculty of Medicine, Tehran University of Medical Sciences, until 2003, and then with the Faculty of Biomedical Engineering, Amirkabir University of Technology, since then, where he is an associate professor of physiology. His research interests include modeling of the biological systems, theoretical neuroscience, and bioinformatics.

# 1 Cell Biology

## 1.1 INTRODUCTION AND OVERVIEW

To understand the background of bioinformatics, including the data itself and the engineering methods involved, one must be familiar with the basics of cell biology. However, the cell biology field is very extensive, and so this chapter presents a review of the main concepts for the beginning bioinformatics student. For the interested reader, specific books and articles cover the different branches of cell biology in a wider scope.

This chapter begins with an explanation of cell structure, separated into the three main cell components: plasma membrane, cytoplasm, and nucleus. This is followed in Section 1.3 by a discussion of proteins, and their use by the cell to perform various functions. Section 1.4 explains the structures and functions of genes, which are the inherited material of the cell. Finally, since knowledge of the cell cycle is crucial to understanding cell behavior, the last section of this chapter is dedicated to this issue. The chapter concludes with review questions.

## 1.2 AN INTRODUCTION TO CELL STRUCTURE

Each cell is composed of three main parts: plasma membrane, cytoplasm, and nucleus.

### 1.2.1 PLASMA MEMBRANE

The plasma membrane forms the boundary between the interior and the exterior of each cell. It is composed of a lipid bilayer, in which proteins and carbohydrates are embedded irregularly to create a fluid-mosaic architecture.

#### 1.2.1.1 Membrane Phospholipids

The majority of molecules, in the cell membrane are lipids, and most are in the form of phospholipids. Phospholipids typically consist of a phosphate group, glycerol, and two fatty acid chains.

An important characteristic of phospholipids is that they possess both a polar head (phosphate) and two nonpolar tails (fatty acid chains). This causes the head to be hydrophilic (water-loving) and the tails to be hydrophobic (water-fearing). When phospholipids make contact with water, the heads are drawn toward the water, and the tails move away. This creates a bilayer structure in the cell membrane. The phospholipid heads point toward the interior and exterior of the cell, where water is closest; the tails are therefore positioned in the center of the membrane, at the furthest point from the water.

#### 1.2.1.2 Proteins

Whereas the lipid bilayer determines the basic structure of biological membranes, proteins are responsible for most membrane functions, serving as specific receptors, enzymes, transporters, and so on (these functions will be discussed later in this chapter). Many membrane proteins extend across the lipid bilayer and are therefore called "transmembrane" proteins. Other proteins that do not span the lipid bilayer are attached to either side of the membrane and are typically bound in place by noncovalent interactions with the transmembrane proteins. Like lipid molecules, many membrane proteins are able to diffuse in the surface of the membrane; however, they are usually unable to move across the membrane (i.e., "flip-flop"). Some proteins on the cell surface are attached covalently with oligosaccharide chains, forming a sugar coating, which is thought to protect the cell surface from mechanical and chemical damage. This coating is also involved in interactions between cells.

The membrane itself has several key characteristics:

1. Fluidity
2. Stability
3. Capacitive property
4. Surface for protein attachment
5. Control of passage of different substances

### 1.2.1.3   Fluidity

Since the molecules in the membrane can diffuse laterally in two dimensions, the membrane lipid can be considered a two-dimensional fluid medium.

### 1.2.1.4   Stability

Although the membrane is a two-dimensional fluid, it is solid in the third dimension, where the third dimension is the direction perpendicular to the membrane surface. Since the membrane is solid in this one dimension, it is reasonably stable; the molecules of the membrane are unable to move across it.

### 1.2.1.5   Capacitive Property

The hydrophobic center of the lipid bilayer prevents the diffusion of ions across the cell membrane. Hence, it is said to possess capacitive properties, comparable to a capacitor in an electrical circuit.

### 1.2.1.6   Surface for Protein Attachment

Proteins are attached to the membrane according to their hydrophobic and hydrophilic properties. Those proteins that are purely polar are *peripheral*; others are *integral*, i.e., embedded in the membrane because they contain both polar and nonpolar components. The polar components of integral proteins are located adjacent to the polar region of the lipid bilayer. Similarly, the nonpolar components are located in the nonpolar region of the bilayer.

### 1.2.1.7   Control of Passage of Different Substances

Lipid-soluble molecules can pass through the membrane by simply dissolving. However, the membrane blocks the simple passage of ionic and polar molecules.

## 1.2.2 CYTOPLASM

The contents of a cell contained within its plasma membrane are called *cytoplasm*. Note that in eukaryotic cells, which contain a distinct nucleus separated by a nuclear membrane, this nucleus is not considered part of the cell's cytoplasm. Cytoplasm contains *organelles*: various membrane-enclosed compartments with a distinct structure, macromolecular composition, and function. Some important organelles found in a typical cell are as follows.

### 1.2.2.1 Mitochondria

Mitochondria are responsible for oxidative phosphorylation reactions; in essence, the cell's respiration. These reactions produce *adenosine triphosphate* (ATP). ATP is effectively the cell's energy currency, spent on various cell activities. Mitochondria are therefore abundant in cells with high-energy expenditure requirements, such as heart muscle cells.

### 1.2.2.2 Endoplasmic Reticula

Endoplasmic reticula are found in the cytoplasm of all eukaryotic cells. They are extensive, labyrinthine networks that fold, translate, and transport proteins and also produce lipids.

### 1.2.2.3 Golgi Apparatus

Found in most eukaryotic cells, the Golgi apparatus organelle's main purpose is to process the proteins and lipids generated by the cell and prepare them for secretion. It has multiple other functions, including the production of proteoglycans to form the extracellular matrices in animal cells.

### 1.2.2.4 Lysosomes

Lysosomes are organelles found in eukaryotic cells that contain digestive enzymes. These enzymes are most active at relatively low pH levels; hence, lysosomes have an acidic interior protected by an external membrane.

### 1.2.2.5 Ribosomes

Found in all cells, ribosome complexes consist of ribonucleic acids (RNAs)* and proteins. Ribosomes are crucial to protein synthesis; by providing a place of interaction for messenger RNA (mRNA) and transfer RNA (tRNA), they help translate mRNA to build polypeptide chains. mRNA contains a "recipe" for building a specific protein, and tRNA transfers amino acids. When a tRNA is attached to mRNA, a new amino acid is added to the protein chain that is currently being synthesized.

## 1.2.3 NUCLEUS

The nucleus organelle is found only in eukaryotic cells and contains the cell's genetic material in the form of chromosomes. This genetic material is protected by a two-membrane envelope that surrounds the nucleus, preventing free diffusion of large

---

* Different types of RNA will be discussed later in this chapter.

molecules and separating chromosomes from cytoplasm. Such separation is vital to prevent mutation and corruption of DNA. Mutation in this sense refers to a permanent change in the DNA, which may alter a trait, or manifest as disease. Small molecules are still able to move in and out of the nucleus through a number of nuclear pores on the envelope surface. These nuclear pores enable the cell DNA to send out information in the form of messenger RNA (mRNA), which is able to move through the pores. This mRNA will later be used in protein formation, a process that will be explained later in this chapter.

## 1.3 PROTEINS AS TOOLS OF THE CELL

### 1.3.1 STRUCTURE OF PROTEINS

Proteins are macromolecules consisting of amino acids and are crucial to every function and process within a cell. Protein structure is of particular importance, as it dictates the protein's biological function. Different proteins have different structures, and these structures may change while the protein performs some action or because of environmental factors such as pH, temperature, and cell membrane voltage.

#### 1.3.1.1 Higher Levels of Protein Organization

A protein's structure can be considered as having four distinct levels: primary, secondary, tertiary, and quaternary. The *primary structure* of a protein refers to its amino acid sequence. The *secondary structure* is the local three-dimensional folding of the polypeptide chain in a short subsequence of amino acids. Secondary structure is defined by two substructures: alpha helices and beta sheets. Alpha helices consist of amino acids arranged in a helical structure, and beta sheets comprise two or more parallel polypeptide chains joined by hydrogen bonds. The *tertiary structure* is the three-dimensional structure of the protein molecule, according to the spatial conformational relationships between the secondary structures. The *quaternary structure* refers to an association of multiple protein or polypeptide subunits into a multiunit protein. Note that not all proteins possess a quaternary structure.

A protein typically has a unique native shape, or native conformation, comprising its secondary, tertiary, and quaternary structures. Once the protein's primary structure—the sequence of amino acids—folds into its native shape, the protein is able to perform its specific function. In most cases, this folding occurs through noncovalent forces.

### 1.3.2 FUNCTIONS OF DIFFERENT PROTEINS

Each protein within the body has a specific function, and proteins can be classified accordingly. Since proteins are an integral part of virtually all cell functions, the range of categories is wide; they include structural proteins, transport proteins, receptors, immunologic proteins, blood carrier proteins, and enzymes.

#### 1.3.2.1 Structural Proteins

Structural proteins are involved in the basic mechanical functions and structure of the cell. For example, ankyrin and spectrin are found in the red blood cell membrane and cause the cell's biconcave shape. Ankyrin defects are the most common cause of hereditary spherocytosis, which reduces the lifetime of red blood cells because of destruction of the cell membrane. This disease causes anemia.

### 1.3.2.2  Transport Proteins

Transport proteins are responsible for the transport of substances across the cell membrane. They can be divided into two types: channels and transporters (carriers).

Transmembrane channel proteins contain a pore, which may be open or closed. Some channel proteins, known as leak channels, are always open and allow a specific ion to pass through. Other channels have gates, which may be open or closed, and regulate the passage of certain ions. There are three types of gated channels:

1. Channels with a voltage sensor, whose gate opens at a specific voltage. An example is a sodium voltage-dependent channel in nerve cells, which initiates electrical impulse propagation.
2. Channels with a ligand site. A ligand in this case is a molecule that binds to a specific site and determines the status of the channel gate (i.e., open or closed). An example is the NMDA receptor in synapses, which is opened by binding to glutamate.
3. Channels regulated by mechanical force. These channels exist in the human ear and open upon receiving sound waves.

Transporter proteins, unlike channels, have no permanent pore. They develop temporary pores upon a change in their conformation, or chemical structure, allowing another specific substance to pass through. This change usually occurs on attachment of another substance to the transporter. An example transporter is the glucose transporter, or GLUT, which controls the entrance of glucose to skeletal muscle cells. Note that the transporter must change its conformation every time a molecule is to be passed through; it reverts to its standard structure between passages.

### 1.3.2.3  Receptors

Receptors are functional macromolecules; in other words, they are large molecules, mainly composed of proteins or glycoproteins, which are capable of some specific function. To perform this function, the receptor must be attached to a specific molecule to which it has affinity—the agonist. The attachment of agonist to receptor alters the chemical structure of the protein, and this conformational shift in turn alters the protein function. In other words, the receptor now gains a specific function, which is either activation of an enzymatic cascade or opening of a channel in the cell.

### 1.3.2.4  Immunological Proteins

Immunological proteins are those proteins critical to the body's defense mechanisms. Three important types of these proteins are antigens, antibodies, and the complement system.

1. Antigens provoke an immune response—i.e., spur the creation of antibodies. They may have entered the body from the outside or have been generated inside the cell (because of infection, for example).
2. Antibodies are produced by *plasma cells* in response to a foreign molecule or invading organism. There are millions of antibody types, each binding to a different antigen. This binding marks the antigen for destruction by the immune system. Antibodies are also needed for complement system activation and phagocytosis.
3. The complement system is a biochemical cascade crucial to the immune system. In an immune response, specific proteins in the complement system are repeatedly cleaved by protease enzymes, creating an activation cascade. This cascade destroys identified antigens by creating numerous pores in their cell membranes; these pores allow molecules to move freely in and out of the cell, leading to its death.

### 1.3.2.5 Blood Carrier Proteins

Carrier proteins are responsible for carrying special substances in extracellular fluids; blood carrier proteins focus on transporting substances in the bloodstream. Examples include serum albumin, which carries hormones, fatty acids, and other molecules; hemoglobin, which carries oxygen and carbon dioxide; and thyroxin-binding globulin (TBG), which carries thyroid hormones.

### 1.3.2.6 Enzymes

Enzymes catalyze (or speed up) a specific chemical reaction. They are vital to cell function, as they allow important processes to occur with sufficient speed and frequency; a speed of a catalyzed reaction may be a million times that of the same reaction without use of an enzyme. An enzyme's name is sometimes derived from its function; for example, lipase is an enzyme which digests the lipids in the small intestine, and protease breaks proteins down into amino acids.

## 1.4 GENES: DIRECTORS OF THE CELL

### 1.4.1 INTRODUCTION TO INHERITANCE

The role of genes in passing traits and characteristics between parents and children is very well known. However, the same genes are also responsible for the continuous functioning of all the cells in the body. To understand how this is possible, we must first study the structure of genes.

### 1.4.2 THE STRUCTURE OF GENES

Living organisms show massive variety in their number of genes, ranging from just a few in some simple cells to 25,000 in a human being. Human genes are arranged sequentially in long double-helix strand molecules, which are called *deoxyribonucleic acids* (DNAs). DNA is composed of *nucleotides*, which themselves are made up of several simple compounds bound together.

The basic components of nucleotides are:

1. *Phosphoric acid.* Each DNA nucleotide has one phosphate group, which may contain one, two, or three phosphates. High-energy nucleotides which are free in the cell have three phosphates, two of which are used during nucleic acid formation. An example is ATP or adenosine triphosphate.
2. *Deoxyribose.* This sugar has five carbon atoms in its circular structure.
3. *Bases.* The five nitrogenous bases are adenine (A), cytosine (C), guanine (G), thymine (T), and uracil (U); the last one is seen only in RNA, where it replaces thymine.

Combined, these three units form nucleotides. There are five possible types of nucleotide, corresponding to the five bases, though only four are used in DNA (as uracil occurs solely in RNA). These four bases—A, C, G, and T—are the four simple symbols that encode all the biological messages and instructions contained in DNA.

A DNA molecule has two longitudinal backbones, one for each strand and each composed of sugar-phosphate groups. These two backbones wind around each other to form a double helix, with one complete turn every 10 base pairs. The A, C, G, and T bases attach to these backbones, and hydrogen bonds form between bases attached to one strand's backbone and those attached to the other. The base pairings are always the same; "A" pairs with "T" using two hydrogen bonds, and "G" pairs with "C" using three hydrogen bonds. Thus, each strand of a DNA double-helix molecule contains a sequence of nucleotides that is exactly complementary to the nucleotide sequence of its partner strand. The hydrogen bonds between base pairs are strong and serve to stabilize the double-helix structure. The linear sequence of nucleotides in the DNA molecule determines the linear sequence of the amino acids in a protein.

Surprisingly, only a small portion of the DNA in a gene actually codes for protein; the rest consists of long sequences of noncoding DNA. The coding sequences are called *exons*, and the interleaving noncoding sequences are called *introns*. Introns are removed during the RNA splicing process, which will be discussed later in the chapter.

### 1.4.2.1 Chromosomes and Chromatin

Chromosomes are found inside both prokaryotic and eukaryotic cells, and act as organized structures to contain the cell's DNA and the proteins which control its functions. In humans, each contains two copies of each chromosome, one inherited from the father and the other from the mother. These paired chromosomes are called homologous chromosomes (homologs). Humans have 46 chromosomes in total, formed by 22 homologous pairs, and one chromosome pair that encodes gender. This final pair may be homologous (in the female case, with two X chromosomes) or nonhomologous (in the male case, with one X and one Y chromosome). The display of the 46 human chromosomes at mitosis is called the human karyotype.

In eukaryotes, such as human cells, DNA is located within the cell nucleus. This presents an interesting issue: when stretched out, the DNA of human cells is almost 2 m in length, and yet can be contained in the nucleus of a human cell, which is over 200,000 times smaller in size. This is possible because of chromatin, a DNA-protein complex which folds and packs DNA molecules into a much more compact structure. This simultaneously strengthens the DNA. There are two basic classes of chromatin proteins: histones and nonhistones. The central part of every nucleosome is composed of eight histone proteins that DNA winds around like a spool. These DNA-histone complex nucleosomes are joined by linker DNA; this has the effect of compressing the pure DNA into its more condensed arrangement. These compacted nucleosome groups are packed together into a short chromatin fiber.

### 1.4.2.2   Genetic Code

The genetic code is the set of rules that relate genetic information encoded by the four DNA or RNA bases to proteins formed by the 20 amino acids. Triplets of nucleotides on mRNA are called *codons*, and there are 64 possible codon combinations. Each codon defines one amino acid, but an amino acid may be coded by several codons; this redundancy helps compensate for random mutations. Table 1.1 gives the RNA codons for the 20 common amino acids. Note that one specific codon represents a signal to start manufacturing a protein molecule, and three codons represent stopping the manufacturing of a protein. The use of three stop codons is because of the importance of correct sequence termination, to prevent the generation of potentially harmful substances.

#### EXAMPLE 1.1

If a section of DNA, reading from left to right has the following genetic code:

   Strand 1 of DNA: ACACCAAGAGGATTAAGG

   1. Then, what will be the sequence of strand 2?
   2. What will be the codon sequences? Suppose that strand 2 is being transcribed.
   3. Use Table 1.1 to determine the amino acid sequence of the new forming polypeptide.

Answer:

   1. Strand 1: A C A C C A A G A G G A T T A A G G
      Strand 2: T G T G G T T C T C C T A A T T C C
   2. A C A C C A A G A G G A U U A A G G
   3. Based on the table, we will see these amino acids in the polypeptide:

      threonine-proline-arginine-glycine-leucine-arginine

### 1.4.3   DNA REPLICATION

DNA replication, as the name suggests, is the process of duplicating a double-helix DNA molecule. This duplication has three steps: (1) initiation, (2) elongation, and (3) termination.

### 1.4.3.1   Initiation

To begin the duplication process, specific initiator proteins attach to the *replication origin*; a particular sequence in the DNA. The two strands of the DNA molecule are separated at the origin site by the enzyme *helicase*, which breaks the hydrogen bonds

**TABLE 1.1**

**RNA Codons for Amino Acids and for Start and Stop**

| | U | C | A | G | |
|---|---|---|---|---|---|
| U | UUU Phenylalanine (Phe) | UCU Serine (Ser) | UAU Tyrosine (Tyr) | UGU Cysteine (Cys) | U |
| | UUC Phe | UCC Ser | UAC Tyr | UGC Cys | C |
| | UUA Leucine (Leu) | UCA Ser | UAA STOP | UGA STOP | A |
| | UUG Leu | UCG Ser | UAG STOP | UGG Tryptophan (Trp) | G |
| C | CUU Leucine (Leu) | CCU Proline (Pro) | CAU Histidine (His) | CGU Arginine (Arg) | U |
| | CUC Leu | CCC Pro | CAC His | CGC Arg | C |
| | CUA Leu | CCA Pro | CAA Glutamine (Gln) | CGA Arg | A |
| | CUG Leu | CCG Pro | CAG Gln | CGG Arg | G |
| A | AUU Isoleucine (Ile) | ACU Threonine (Thr) | AAU Asparagine (Asn) | AGU Serine (Ser) | U |
| | AUC Ile | ACC Thr | AAC Asn | AGC Ser | C |
| | AUA Ile | ACA Thr | AAA Lysine (Lys) | AGA Arginine (Arg) | A |
| | AUG Methionine (Met) or START | ACG Thr | AAG Lys | AGG Arg | G |
| G | GUU Valine Val | GCU Alanine (Ala) | GAU Aspartic acid (Asp) | GGU Glycine (Gly) | U |
| | GUC (Val) | GCC Ala | GAC Asp | GGC Gly | C |
| | GUA Val | GCA Ala | GAA Glutamic acid (Glu) | GGA Gly | A |
| | GUG Val | GCG Ala | GAG Glu | GGG Gly | G |

in the double-helix structure to form a replication fork. The fork consists of two strands: the leading strand, and the lagging strand, which run in opposite directions. Single-strand binding proteins (SSBPs) then bind to newly formed single strands and prevent their reattachment. The enzyme primase catalyses the formation of primers, short segments of nucleic acid sequences, on the template DNA. The creation of these primers allows the next step, elongation.

### 1.4.3.2 Elongation

In this step, *DNA polymerase* attaches to the primers and begins to polymerize a complementary strand for the DNA single strand. DNA polymerase can only synthesize new DNA from the 5′ to the 3′ direction of the newly formed DNA (which is equivalent to the 3′ to 5′ direction of the template DNA).

Understanding the concepts of "leading strand" and "lagging strand" are necessary to understand replication. Remember that two strands run in opposite directions:

5' to 3' and 3' to 5', respectively. Then, the leading strand, with its 3' OH oriented toward the fork, is elongated simply by appending nucleotides to this end. However, DNA polymerase cannot append new nucleotides to the 5' end of a growing chain, creating a problem for the lagging strand. Okazaki solved this problem by showing that the lagging strand is replicated by treating it as a series of short segments, called Okazaki *fragments*. These are synthesized in the normal 5' to 3' direction in the new strand and later connected.

### 1.4.3.3   Termination

Termination, the final step, occurs in two scenarios: either the replication forks collide or reach the end of a linear DNA molecule.

### 1.4.4   TRANSCRIPTION

Since DNA is contained in the nucleus of a eukaryotic cell, and most cell functions are performed in the cytoplasm, there must be some means for DNA genes to control the cytoplasm's chemical reactions, specifically the generation of proteins. This is achieved by another nucleic acid, RNA, formed via the *transcription* process. Unlike DNA, this RNA is able to move through the pores in the nucleus membrane into the cytoplasm, where it controls protein synthesis. Like DNA, RNA is constructed of nucleotides, complexes of phosphate groups, sugars, and vases. However, there are a few key differences. First, an RNA nucleotide uses the sugar *ribose* rather than deoxyribose. Second, as mentioned earlier in this chapter, the thymine is replaced by another pyrimidine, *uracil*. Also, an RNA molecule consists of only one strand, and its chain of nucleotides is much shorter than the chain found in DNA.

RNA synthesis involves the temporary separation of the two strands of the DNA double helix, so a single strand may be used as a template. The nucleotide sequences in DNA are transcribed, or recoded, into sequences in messenger RNA (mRNA). The particular section of DNA that is transcribed into RNA is a transcription unit. More specifically, the nucleotide triplets in the DNA are transcribed into mRNA codons. These codons describe the sequence of amino acids required for a protein to be synthesized in the cytoplasm; a sequence of mRNA codons effectively acts as a protein "recipe." Note that the term "messenger RNA" reflects the fact that mRNA acts as an information carrier between the DNA (which cannot leave the cell nucleus) and the cell cytoplasm.

The transcription process is accomplished under the influence of the enzyme RNA polymerase. As with DNA replication, the process has three steps: initiation, elongation, and termination. Following the brief overall view above, these steps will now be described in more detail.

1. Initiation
    Rather than the primers used to begin DNA replication, transcription begins simply when RNA polymerase binds to the DNA. However, this binding requires a sequence of nucleotides called the *promoter*, which is found at the end of the DNA strand prior to the gene itself. Because of its appropriate complementary structure, the RNA polymerase is attracted to this promoter and attaches to it. The polymerase then unwinds about

two turns of the DNA helix and separates these portions into two separate strands. Only one of these is used as a template for synthesis.

2. Elongation

The polymerase moves along the template strand, temporarily unwinding and separating the two DNA strands and adding new complementary RNA nucleotides to the end of the newly forming RNA chain. Remember the complementary bases: G to C and A to U (as uracil replaces thymine in RNA). By breaking two of the phosphate radicals away from each new RNA nucleotide, the RNA polymerase is able to free up large quantities of energy. This energy allows the nucleotide's remaining phosphate to form a covalent bond with the ribose on the end of the growing RNA molecule.

3. Termination

At the end of the DNA gene is a final sequence of DNA nucleotides called the *chain-terminating sequence*. When the polymerase recognizes this sequence, it breaks away from the DNA strand, to be reused in the RNA replication process (either for the current DNA strand or a different one). At the same time, the new RNA strand detaches from the DNA template and is released into to the nucleoplasm.

After transcription, the new RNA molecule contains a great many redundant nucleotide sequences, approximately 90% of its total length! At this stage, it is referred to as pre–messenger RNA. The process of *RNA splicing* removes these unnecessary parts—the introns—and joins together the remaining vital sections—the exons.

*Open question*: Investigate which of the two strands of DNA is used as a template for mRNA synthesis. Why is the alternative strand not used for this purpose?

### 1.4.4.1 Three Kinds of RNA

There are three types of RNA, each of which plays an important role in protein formation. These are:

1. *Messenger RNA (mRNA)*, which carries protein "recipes" from the DNA to the cell cytoplasm to control protein formation.
2. *Transfer RNA (tRNA)*, which transports amino acids to the ribosome to be used in protein molecule construction.
3. *Ribosomal RNA (rRNA)*, which forms the ribosome (combined with various proteins)

Two of these—mRNA and tRNA—have already been encountered in this chapter. It is important to remember the functions of these different types.

### 1.4.4.2 tRNA Anticodons

Now that mRNA has been explained in more detail, let us examine tRNA. tRNA transfers amino acids to the ribosomes for protein synthesis; to achieve this, each type of tRNA combines with one specific amino acid, attaching it to one end. Once the tRNA reaches the ribosomes, it delivers its amino acid to the appropriate place in the newly forming protein by identifying the matching codon on the mRNA. The specific portion of tRNA that allows it to recognize the correct codon is a triplet of nucleotide bases called an *anticodon*. During protein formation, the anticodon bases form hydrogen bonds with the codon bases of the mRNA, thus ensuring the correct amino acid sequence in the final protein.

### 1.4.5 TRANSLATION

Translation is the process of decoding mRNA into the ribosome to produce a specific protein, according to the established mappings between codons and amino acids. When an mRNA contacts with a ribosome, it travels through it, and as it travels, its codons are translated by the ribosome and tRNA into specific amino acids. The resulting sequence of amino acids forms a polypeptide chain or protein. When an mRNA "stop" codon reaches the ribosome, this terminates the protein molecule. Note that a single mRNA molecule can be used to form protein molecules in several ribosomes at the same time. These ribosomes, bound to the same mRNA molecule form clusters known as *polyribosomes*.

## 1.5 CELL CYCLE

### 1.5.1 CELL CYCLE: AN OVERVIEW

The cell cycle applies specifically to eukaryotes and describes the process by which the cell divides and replicates itself. The cycle is highly regulated and consists of four stages. Arguably the most remarkable characteristic of the cell cycle is the *mitosis* or *M phase* in which chromosomes condense, align, and separate prior to cell division. This is followed by the *G1 phase*, *S phase*, and *G2 phase*. The G1 phase creates important enzymes that are required for the DNA synthesis performed in the S phase. Similarly, the G2 phase produces proteins required for the next round of mitosis (M phase). The M phase, because of its significance, is discussed here in more detail.

### 1.5.2 MITOSIS

Mitosis is the process by which one cell divides into two cells. More specifically, the chromosomes in the cell's nucleus are separated into two identical sets. Note that the replication itself actually occurs in the *S* phase. Mitosis is followed by cytokinesis, which divides the physical components of the cell—the nucleus, membrane, and cytoplasm—to form two daughter cells, both identical to the parent cell.

#### 1.5.2.1 Mitotic Apparatus

Two pairs of centrioles, organelles found in the cell cytoplasm, lie close to each other near one pole of the nucleus. Each pair of centrioles, with its attached dense pericentriolar material, is called a centrosome. Shortly before mitosis, protein microtubules growing between the two centrosomes push the centrosomes apart. This set of microtubule fibers is called the mitotic spindle, and the entire set of microtubules plus the two centrosomes is called the mitotic apparatus.

### 1.5.2.2  Four Stages of Mitosis

1. Prophase

   At the start of the prophase stage, the chromatin in the cell nucleus condenses into chromosomes. Each chromosome contains two identical chromatids, joined together at the chromosome's centromere. At the same time, the centrosomes are pushed toward opposite sides of the nucleus by the microtubules. This begins the *prometaphase* stage, in which the nuclear envelope is fragmented, causing the cell to enter *open mitosis* (as the nuclear envelope has opened). The microtubules pull the paired chromatids in each chromosome centromere in different directions; one chromatid of each pair moves toward one cellular pole.

2. Metaphase

   The chromatids are pulled tightly to the center of the cell, lining up along the equatorial plane between the two poles. The cell then moves into the *anaphase* substage, where the two chromatids of each chromosome are pulled apart at the centromere. All 46 pairs of chromatids are separated to form two daughter separate sets, each containing 46 chromosomes.

3. Telophase

   The telophase concludes mitosis by disassembling the mitotic apparatus and forming a new nuclear membrane, which develops around each set of daughter chromosomes. The cell then moves into cytokinesis, a separate process that performs physical division of the entire cell, not just the nucleus.

## 1.6  SUMMARY

This chapter reviews the basic concepts in cell biology. The fundamental molecules such as DNA and RNA are briefly explained, along with their roles in encoding proteins. The processes of replication, transcription, and translation are described, concluding with a description of the cell division cycle.

## 1.7  PROBLEMS

1. Define each of these concepts in a short phrase.

   Base, Chromosome, Deoxyribose, DNA, DNA bound proteins, Double helix, Enhancer, Enzymes, Exone, Gene, Gene Expression, Genetic code, Histone, Introne, mRNA, Nuclear membrane, Nucleic Acid, Nucleosome, Nucleotide, Nucleus, Promoter, Replication, Ribose, RNA, RNA polymerase, Transcription, Translation, tRNA, Inducer, Mutation, Nucleotide regulatory protein, Ribosome.

# 2 Bioassays

## 2.1 INTRODUCTION AND OVERVIEW

In this chapter, we will briefly introduce bioassays, mechanisms through which practitioners of bioinformatics collect much of their data. Being familiar with these bioassays will help the reader to understand the bioinformatics field better. Some of the more important bioassays we will discuss are Southern blotting, fluorescent in situ hybridization, the Sanger (dideoxy) method, and polymerase chain reaction (PCR). Finally, we will study the methods of analyzing protein structure and function as well as methods of studying gene expression and function.

## 2.2 SOUTHERN BLOTTING

The Southern blot is a method for detecting a given DNA signature within a series of samples. For example, if one knows the sequence of a particular gene, he/she might use a Southern blot to examine DNA samples from a population of organisms to identify which individuals contained a copy of the reference gene.

The Southern blot procedure requires the mixing of the DNA samples with the reagents that will cut the DNA fragments into relatively small chains. These fragments, one per strain, are loaded into an electrophoresis gel. Once the fragments have been sorted by size, they are transferred—or blotted on—to another medium. The blot is then heated to fix the position of the DNA fragments. A marked probe, often an RNA strand formed to be complementary to the seeked DNA signature is washed over the blot and is allowed to bind to the single-stranded DNA fragments. After the residual probe is rinsed away, the blot is read to investigate which lanes contain the probe. These are the lanes corresponding to the target samples or organisms.

Many different variants of the Southern blot exist. The Northern blot detects probes within RNA blots instead of DNA blots; the Western blot detects probes within protein blots instead of DNA blots and Southwestern blots detect DNA-binding proteins.

## 2.3 FLUORESCENT IN SITU HYBRIDIZATION

Fluorescent in situ hybridization (FISH) is another method to detect probe sequences of DNA among the entire chromosomes. Briefly, the probe is constructed from DNA that has been tagged with fluorescent markers. The chromosomes are affixed to some medium and then allowed to hybridize with the probe. Then the excess probe is washed away from the environment. Finally, a fluorescent microscope is applied to observe where the bright species are detected.

FISH is particularly useful when searching for known markers of genetic diseases. The probe should reflect the allele associated with disease, and the microscopy would reveal whether this allele is detected on the target chromosome. FISH also useful to investigate the prevalence of particular genes in a population.

## 2.4   SANGER (DIDEOXY) METHOD

The Sanger method, invented in 1975, is a technology to sequence (enumerate, in order, all of the base pairs within a sample of) DNA.

Because the position of each nucleotide type (A, C, G, T) is to be determined independently, the DNA sample is divided into four batches. Each batch incorporates its portion of the DNA sample, fluorotagged nucleotides, and exactly one type of dideoxynucleotide. The four dideoxynucleotides are able to participate in the replication of the unzipped DNA strand (so there is one dideoxynucleotide analog for each A, C, G, and T). However, dideoxynucleotide have one important difference: because they lack the chemical subunits to continue the chain, the replication ends as soon as a dideoxynucleotide binds to the sequence. As a result, the fluorescent nucleotides are added in greater concentration, so as to allow fragments of variable length to arise.

Once the four reactions have run for sufficient time, the products are denatured and then applied to a gel. On the gel, each dideoxynucleotide is contained within a single lane. As the copied fragments diffuse through the gel, they are effectively sorted by size from the longest to the shortest. If the proportions are selected carefully, and the reaction had run long enough, there should be representatives of many different fragment lengths in the sample; most are different in length by only one nucleotide. Reading the results off the gel allows the researcher to reconstruct the sequence of nucleotides: if the nucleotide at position $n$ is G, then the ddGTP dideoxynucleotide lane should have a strong signal at the $n$ base pair distance; no other lane should have a signal at that length because all other fragments that start at the beginning of the DNA strand should have terminated with ddGTP, or they should have bound a fluorescent G to that location and continued elongating (forcing them to be deposited at another gel location).

Because the chain-terminating residues are dideoxynucleotides, the method is sometimes also referred to as the dideoxy method.

## 2.5   POLYMERASE CHAIN REACTION

Polymerase chain reaction (PCR) is a method for amplifying a portion of a DNA sequence. Starting with a small sample, PCR is able to generate millions of copies that are suitable for additional analysis. PCR is widely used throughout the life sciences.

PCR consists of multiple cycles of varying temperature. Within each cycle, the DNA present in solution is heated until it melts (separates into individual strands); at a lower temperature, the primers—sequences of DNA that are specific to the starting location that you wish to copy—are allowed to bind to the single strands of DNA; the temperature is then changed so that the DNA polymerase can act, adding complementary nucleotides to the single DNA strand. The cycles are repeated, until a reasonable amount of DNA is obtained. Note that the maximum gain per cycle is 100%, so each cycle will no more than double the total amount of the target DNA present in the original solution. Ten cycles increase the amount of DNA by less than or equal to 1,204 times.

## 2.6 ANALYZING PROTEIN STRUCTURE AND FUNCTION

### 2.6.1 X-ray Crystallography

X-ray crystallography is used to infer the consensus conformation (three-dimensional structure) for proteins. Its requirements are strict: the protein must be able to form a very regular crystal with few impurities. The crystal is exposed to X-rays, and sensors record the diffraction pattern. Software is used to interpret the diffraction pattern into a three-dimensional charge-density map of the unit cell that can (hopefully) be translated into a geometric arrangement of amino acids.

### 2.6.2 NMR Spectroscopy

Nuclear magnetic resonance (NMR) spectroscopy can be a suitable method to obtain three-dimensional structural information about relatively small proteins, particularly those that defy crystallization. The target protein must be labeled with C13 and N15 before NMR spectroscopy is viable because these atoms have spins that best support interpretable data coming out of the method. The labeled proteins are purified and then subjected to a strong magnetic field. The interaction between the spin of the key nuclei and the magnetic field produces a signal that is recorded. To ensure that there are enough data to infer the protein structure, multiple dimensions of data must be captured. The multidimensional signals are sent to a software tool that will then return an estimated model of the location of the atoms within the protein.

### 2.6.3 Affinity Chromatography

Affinity chromatography is a method to perform a number of separation steps on proteins. For example, it can be used to identify proteins that bind to specific antibodies, or it can be used to identify proteins that bind another specific protein. Can also be used to identify proteins that are receptive to specific ligands.

In all these cases, the fixed element—be it an antibody, protein, or ligand—is affixed to a solid medium, often a bead, and then these solid media are packed into a column so as to that liquids can pass through. The solution that contains the candidate proteins, usually obtained directly from live cytoplasm, is poured through the column. Excess solution is washed out of the column. At the final step also called elution, the solution washes the fixed proteins from the column. The final wash product can be analyzed to determine what species were bound to the solid medium.

### 2.6.4 Immunoprecipitation

Immunoprecipitation is in principle similar to affinity chromatography, except that it is designed to identify protein antigens that bind to a given antibody. The target antibody is affixed to a solid medium, and then a solution containing multiple proteins is introduced. Once washed, the final set of protein antigens, bound to the antibodies, constitute the desired end product.

### 2.6.5   Two-Hybrid Systems

This mechanism is designed to test for a particular protein-protein or protein-DNA interaction using a reporter gene. The premise is that if the two reacting species are hybridized, each with one portion of the activation complex for the reporter gene, then any evidence of the reporter gene being expressed suggests that the two halves of the activation complex were joined. That is, the protein-protein or protein-DNA pair must have interacted for the two halves of the activation complex to be bound together sufficiently for the downstream (reporter) gene to have been expressed.

### 2.6.6   Phage Display Method

This method detects protein-protein interactions. It involves introducing genes into a virus that infects the *Escherichia coli* bacterium. The DNA encoding part of the protein of interest is fused with a gene encoding part of one of the proteins that forms the viral coat. When this virus infects *E. coli*, it replicates, as a result produces phage particles that display the hybrid protein on the outside of their coats. This bacteriophage can then be added to a large pool of potential target proteins. The protein that binds to the new bacteriophage envelope protein can then be captured and assessed.

### 2.6.7   DNA Footprint Analysis

DNA footprint analysis is used to identify where proteins bind on DNA strands. Starting from a collection of copies of a candidate DNA fragment—amplified via PCR, and fluorotagged—the target protein is added, and allowed time to bind. An enzyme that can cut the DNA strands is then added to the mixture and is given just enough time to cleave the DNA strands once. The solution is then introduced to a gel and allowed to separate alongside another solution of the same DNA that was not allowed to bind to protein (but was also cut into fragments). The two gel spectra should be identical *except* for a single band, where the protein that was bound to the parent DNA strand did not allow the enzyme to cleave. This gap is referred to as the footprint.

### 2.7   STUDYING GENE EXPRESSION AND
### FUNCTION—DNA MICROARRAYS

DNA microarrays have become a preferred high-volume method for detecting gene expression levels. A DNA microarray itself is a matrix of wells in which each well is impregnated with a probe. A sample is introduced to the microarray, and those wells in which the sample binds to the probe are detected. Even though the probes must be designed specifically for the sample that will be analyzed, there are standard microarray templates that help encourage reuse.

To monitor gene expression levels, for example, the probes might be constructed from DNA correlated with key genes. As the mRNA sample is washed over the wells, those cDNA-mRNA pairs that bind will allow the reader to identify the expression level of the genes encoded on the microarray. As useful as this sounds, it should be clear that using the same design for a microarray twice will allow for comparisons between expression levels in that organism. This may be done to monitor the progress of a disease, estimate the impact of therapeutic interventions, or collect time series data on the natural variability of gene expression levels.

Typically, the probes have fluorescent tags attached to them that are activated only, then the probe binds to some component in the sample. The microarray is read by taking an image of the array and/or scanning the array, and correlating the brightness of the image with the known locations of the wells and inferring strength of activation. To help discriminate between true positives and false positives, microarrays are often designed with multiple probe variations. Commercial microarray systems include software to help interpret the results using statistical methods.

### 2.7.1 SPOTTED MICROARRAYS

A spotted microarray is one in which the probes are manually assigned locations on which they should be deposited. This type of system provides flexibility to the researcher, with a somewhat reduced cost from purchasing the large, preordained oligonucleotide microarrays.

### 2.7.2 OLIGONUCLEOTIDE MICROARRAYS

An oligonucleotide microarray is one in which the probes are built into the wells, instead of simply being deposited onto the medium at a known location. As a result, these are almost always built off-site, though they can be custom-specified.

### 2.7.3 SNP MICROARRAYS

These are used to identify genetic variation in individuals. Short oligonucleotide arrays can be used to identify the single nucleotide polymorphisms (SNPs) that are thought to be responsible for genetic variation and the source of susceptibility to genetically caused diseases. These SNP microarrays are also being used to profile somatic mutations in cancer.

## 2.8 SUMMARY

In this chapter, the most popular molecular measurement methods used in molecular biology were reviewed. The objective of the chapter was not to give a comprehensive understanding about each method; rather, it was to give a brief review of each system. Readers are encouraged to use more detailed text on biological assays to further their knowledge on the high throughput systems.

## 2.9   PROBLEMS

1. Describe the function and usage of different types of DNA microarrays.
2. Explain the main steps of PCR.
3. Under what conditions would NMR spectroscopy be preferred over X-ray crystallography?
4. What is the difference between a Southern blot and a Northern blot?
5. What does "NOE" stand for (with respect to NMR spectroscopy)? What advantages does it provide?
6. What is the purpose of PCR?
7. How does the Sanger method sequence DNA?

# 3 Review of Some Computational Methods

## 3.1 INTRODUCTION AND OVERVIEW

In this chapter, some methods commonly used in bioinformatics and systems biology are reviewed. In order not to assume a previous background in some fields of computational sciences, this chapter begins with some fundamental concepts of these theories and gradually approaches the more advanced concepts.

## 3.2 INTRODUCTION TO PROBABILITY AND STOCHASTIC PROCESSES THEORIES

A large group of the phenomena, systems, and techniques used in bioinformatics and systems biology are "deterministic." These phenomena are referred to as deterministic because every time these phenomena are observed and evaluated, the same values are observed. For instance, the result of $3 + 4$ is always 7 regardless of when and how this summation is made. Similarly for signals and time series, if every time a time series is recorded/registered and the exact same values are obtained, the time series is called to be deterministic. However, many biological quantities and time series are not deterministic, and each time we measure/observe them, we obtain different results. In fact, the majority of biological quantities and time series are "probabilistic" or "stochastic." A probabilistic phenomenon/variable, $x(w)$, is a variable that gives a potentially different value in each measurement. The variable $w$ emphasizes the fact that each measurement of the variable $x$ can result in a different value. Probabilistic processes can be either time-dependent or time-independent. If a random variable is time-dependent, it is often referred to as a "stochastic process." A stochastic process, although statistically similar in all recordings, contains some stochastic variations in each of the recordings. A stochastic process, $x(w,t)$, is typically represented by two variables; the first available $t$ represents time, and the second variable $w$ identifies a particular outcome of the stochastic process. Since a stochastic process is a more general formulation of a probabilistic process (due to time dependency), in the formulations below of probabilistic systems, we focus on stochastic processes. In addition, a stochastic process can be either continuous time or discrete time. In this chapter, we will provide the formulation for both continuous and discrete stochastic processes.

An interesting stochastic process is the "white noise." White noise is defined as a stochastic process in which the value of the signal at each time has absolutely no statistical dependency on the value of the signal at any past or future times. A discrete-time example of such a process is a time series of numbers obtained in consecutive tosses of a fair coin. In such a process, the outcome of each toss does not depend on the past or future outcomes. The study of the white noise and the techniques to detect or remove it from a signal is a dynamic field of research. DNA microarray data often contains a significant amount of white noise, and therefore, any processing of such data requires the consideration and modeling of this noise.

### 3.2.1 METRICS FOR STOCHASTIC PROCESSES

At this point, we are to make our notation slightly simpler and shorter. We drop the variable $w$ from the notation and use $x(t)$ to represent $x(w,t)$; this makes the notation easier. However, it is necessary not to confuse $x(t)$ with a deterministic signal and keep in mind the random nature of the process.

The first metric to quantify the properties of a stochastic process is the probability density function.

#### 3.2.1.1 Probability Density Function

As mentioned above, the difference between a deterministic signal and a stochastic process is the fact that once we have one recording of a deterministic signal, we know everything about it, while every measurement of a stochastic signal is merely one outcome of a random sequence, and therefore, this outcome cannot represent all aspects of the stochastic process. The question then becomes how a stochastic process can be described. The solution is simply using a probability function to express the likelihood of having a value for a signal at any time $t$. The probability density function (PDF) of a process $x(w,t)$ is represented as $p_X(w,t)$ or simply $p_X(t)$.

This probability function can be estimated using either the conceptual definition of the process or the measurements made from the process. For instance, consider the example of tossing a fair dice. To identify $p_X(1,t)$, $p_X(2,t)$, $p_X(3,t)$, $p_X(4,t)$, $p_X(5,t)$, and $p_X(6,t)$, one can easily say that since the dice is "fair" the likelihood of occurrence of each of the outcomes is the same, and therefore,

$$p_X(1,t) = p_X(2,t) = p_X(3,t) = p_X(4,t) = p_X(5,t) = p_X(6,t) = 1/6 \qquad (3.1)$$

A practical method of estimating PDF is using the prior recordings of the outcomes of the process. For instance, for a dice (which may or may not be a fair dice), we can simply toss the dice many times while recording the outcomes and then divide the number of occurrence of each number $i$, $N(i)$, by the total trials, $N$, to obtain an estimate of the probability, i.e.,

$$p_X(i,t) = N(i)/N \qquad (3.2)$$

PDF is an extremely useful quantity that allows calculation of many more metrics about a process, as discussed next.

### 3.2.1.2   Mean and Variance

As in probability theory, once the PDF is available, it is always desirable to know the average of a stochastic process at any time $t$. In practical applications, where we have many recordings of a stochastic signal, we can average all the available recordings at time $t$ and consider this value as the estimation of $m(t)$, i.e., mean of the signal at time $t$. Mathematically speaking, the actual value of the mean function at all times can be computed as follows:

$$m(t) = E\big(x(t)\big) = \int_{-\infty}^{-\infty} x(t) p_X(t) dx(t) \tag{3.3}$$

The function $E(.)$ is called the expectation function, also referred to as ensemble averaging function. For a discrete stochastic process $x(w_i,n)$ that is defined at time points $n$, the mean function is defined as

$$m(n) = E\big(x(w_i,n)\big) = \sum_{i=-\infty}^{-\infty} x(w_i,n) p_X(w_i,n) \tag{3.4}$$

where $p_X(w_i,n)$ is the probability of having outcome $w_i$ at time $n$.

Often only one average function, i.e., mean, is not sufficient to represent all characteristics of an entire stochastic process. Variance is another popular function that is very effective in expressing the average variations and scattering of the data around the mean. This function is defined as

$$\sigma^2(t) = E\big((x(t) - m(x))^2\big) = \int_{-\infty}^{-\infty} \big(x(t) - m(t)\big)^2 p_X(t) dx(t) \tag{3.5}$$

As can be seen, the variance function is the statistical average of the second-order deviation of the signal from its mean at each time. Similarly, the discrete variance function is defined as

$$\sigma^2(n) = E\big((x(w,n) - m(n))^2\big) = \sum_{i=-\infty}^{+\infty} \big(x(w_i,n) - m(n)\big)^2 p_X(w_i,n) \tag{3.6}$$

### 3.2.1.3   Expectation Function

A closer look at the equations presented above emphasizes the fact that although mean is the statistical (or ensemble) average of $x(t)$, variance is nothing but statistical average of $(x(t) - m(t))^2$. This statistical averaging can be extended for any general function $g(x(t))$, namely,

$$\overline{g}(t) = E\big(g(x(t))\big) = \int_{-\infty}^{-\infty} g(x(t)) p_X(t) dx(t) \tag{3.7}$$

Similarly, for the discrete processes, we have

$$\overline{g}(n) = E\big(g(x(w,n))\big) = \sum_{i=-\infty}^{+\infty} g(x(w_i,n)) p_X(w_i,n) \tag{3.8}$$

Some other popular functions whose expectations are useful in practical applications are "moment functions." The moment of degree $k$ is defined as $E(g(x(t)))$ where $g(x(t)) = x(t)^k$. As can be seen, the mean function is nothing but the moment of degree 1, and variance is closely related to the moment of degree 2. The moment of degree 2 is often considered as the statistical power of a signal.

### 3.2.3 Gaussian Distribution

The PDF can have different distributions rising to random processes with some specific properties. Among these different forms of PDF, Gaussian function has some unique properties that make this distribution the most practically important random distribution. Before describing these properties, let us review the mathematical definition of Gaussian distribution (also known as normal distribution). The PDF for this distribution is defined as follows:

$$p_X(x) = \frac{1}{\sigma\sqrt{2\pi}} \exp\left(-\frac{(x-m)^2}{2\sigma^2}\right) \tag{3.9}$$

As it can be seen, this distribution is completely defined using only two variables; mean and variance. This is one of the interesting properties of Gaussian random variables.

However, what makes Gaussian distribution a very interesting one is the fact that very many random phenomena in nature can be accurately estimated using this distribution. In simpler words, the majority of random variables in nature can be assumed to be normally distributed or can be approximated using a Gaussian distribution. The justification for such a rather radical assumption comes from a famous theorem in probability theory called "central limit theorem." We do not intend to either prove or even formulate central limit theorem here, but it is useful to know that according to this theorem, loosely speaking, when a phenomenon is affected by so many random independent variables with arbitrary PDFs, the distribution of the phenomenon will be Gaussian. Since many random variables in nature are indeed affected by a large number of independent random variables, it is natural to observe that they are distributed according to Gaussian distribution.

### 3.2.4 Correlation Function

To express the time patterns within a stochastic signal as well as the interrelations across two or more stochastic signals, we need to have some practically useful measures and functions. For instance, since almost all bioinformatics time series have some type of periodicity, it is extremely useful to explore such periodicities using techniques. However, since these signals/time series are stochastic, we cannot simply draw determinist conclusions from them directly just by processing only the recording of the signal. The approach we take in this section to address the above issue is rather simple: we construct meaningful determinist signals from a stochastic process and then process/analyze these representative determinist signals using methods such as Fourier Transform. As we will see later in this chapter, these deterministic signals and measures are conceptually interesting and practically very meaningful. Although the definition of these signals and measures can be described for nonstationary processes too, because almost all biomedical applications processes and signals are assumed to be stationary, we focus only on stationary processes and specialize all definitions toward stationary processes.

The first function we discuss here is autocorrelation function. This function calculates the similarity of a signal to its shifted versions, i.e., it discovers the correlation and similarity between $x(t)$ and $x(t - \tau)$, where $\tau$ is the amount of shift. Specifically, the autocorrelation function, $r_{XX}(\tau)$, is defined as

$$r_{XX}(\tau) = \int_{-\infty}^{+\infty} x(t)x(t-\tau)p_{XX}\big(x(t),x(t-\tau)\big)dt \qquad (3.10)$$

In the above equation, $p_{XX}(x(t), x(t - \tau))$ is the joint PDF of $x(t)$ and $x(t - \tau)$. An interesting property of this function is its capability to detect periodicity in stochastically periodic signals. For such signals, whenever $\tau$ is a multiple of the period of the signal, the similarity between the signal and its shifted version exhibits a peak. This peak in the autocorrelation signal can be quantitatively captured and measured. In other words, a periodic autocorrelation signal not only reveals the periodicity of the stochastic process but also measures the main features of this periodicity such as the period of oscillation.

Another feature of the autocorrelation function deals with the value of $\tau$ at which the autocorrelation function reaches its maximum. A simple heuristic observation states that maximum similarly is gained when a signal is compared with itself (i.e., when there is zero shift). This argument explains why the maximum of the autocorrelation function always occurs at $\tau = 0$.

Yet another interesting observation about the autocorrelation function deals with calculating this function for the white noise. From the definition of the white noise, we know that there is no dependence or correlation among random variables in consecutive times. This means that for any value $\tau \neq 0$, the autocorrelation function is zero. We also know that the similarity of a signal to itself is maximal. This means that the autocorrelation function for the white noise is indeed a spike (impulse) function which has a large value at the origin and zero elsewhere.

The definition of the autocorrelation function for a discrete signal $x(n)$ is simply the same as the continuous case except for the substitution of the integral with a summation:

$$r_{XX}(m) = \sum_{n=-\infty}^{+\infty} x(n)x(n-m)p_{XX}\big(x(n),x(n-m)\big) \qquad (3.11)$$

In the above equation, $m$ identifies the amount of the shift in the discrete domain, and $p_{XX}(x(n), x(n - m))$ is the joint PDF of the random variables $x(n)$ and $x(n - m)$.

A natural and logical extension of the above functions is cross-correlation function. This function identifies any relation between a stochastic process $x(t)$ and the shifted version of another stochastic process $y(t)$, i.e., $y(t - \tau)$, as follows:

$$r_{XY}(\tau) = \int_{-\infty}^{+\infty} x(t) y(t-\tau) p_{XY}\big(x(t), y(t-\tau)\big) dt \qquad (3.12)$$

In the above equation, $p_{XY}(x(t), y(t - \tau))$ is the joint PDF of random variables $x(t)$ and $y(t - \tau)$, and $r_{XY}(\tau)$ is the autocorrelation function between the two random variable $x(t)$ and $y(t - \tau)$.

The main application of this function is identifying potential cause-and-effect relationship between the random processes. As a trivial example, assume that we are to discover if there is a relationship between the blood pressure signal $y(t)$ and the ECG recordings of a patient, $x(t)$, measured an hour after the blood pressure was measured, i.e., $\tau = 1$ min. If the autocorrelation function showed a peak at $\tau = 1$ min, then one can claim that there might be some time-delayed relation between the blood pressure and the electric activates of the heart muscles, i.e., ECG. Although this trivial example may seem too simple and obvious, there are so many other factors, such as exercise and nutrition patterns, whose potential effects on the function and activities of the heart are constantly investigated using the cross-correlation function.

The cross-correlation function for the discrete processes is defined as

$$r_{XY}(m) = \sum_{n=-\infty}^{+\infty} x(n) y(n-m) p_{XY}\big(x(n), y(n-m)\big) \qquad (3.13)$$

It is important to note that all autocorrelation and cross-correlation functions defined above, although calculated from random variables, are indeed deterministic signals.

### 3.2.5 STATIONARY AND ERGODIC STOCHASTIC PROCESSES

A subset of stochastic processes provides some practically useful characteristics. This family, referred to as "stationary processes in wide sense," is the one where mean and variance functions remain constant for all time, i.e., $m(t) = m_0$ and $\sigma(t) = \sigma_0$. In other words, although the probability function $p_X(t)$ of such processes can change through time, the mean and variance of the process stay the same at all times. Such a simplifying assumption helps in processing a number of practically useful signals. For instance, it is often the case that the mean and variance of signals such as DNA microarray time series do not change at least during a rather large window of time. Such an observation allows calculation of mean and variance for only one time point because the assumption of stationarity in wide sense states that the mean and variance functions for all times will be the same.

A subset of wide-sense stationary processes are "stationary in the strict sense" processes in which the probability function $p_X(t)$ is assumed to be independent of time. This means that if one finds the PDF for one time step, the same exact PDF can be used to describe all statistical characteristics of the process in all times. From the definition, one can see that while all strict-sense stationary processes are also wide-sense stationary, the opposite is not true. This means that the strict-sense stationary assumption is a stronger assumption and therefore applicable to fewer real applications.

An interesting and very popular family of wide-sense stationary processes is the set of wide-sense stationary Gaussian processes. In such processes, since the PDF has only two parameters, mean and variance, once these two parameters are fixed in time, the PDF becomes the same for all time. This means that for Gaussian processes, strict- and wide-sense stationary concepts are the same. Since many processes can be approximated by some Gaussian process, such stationary Gaussian processes are extremely important in signal processing applications especially bioinformatics. For instance, in almost all processing techniques used in bioinformatics, the stochastic processes are assumed to be stationary Gaussian. Hereafter, since strict- and wide-sense stationary Gaussian processes are the same, in referring to such processes, we only use the word *stationary* without mentioning the details.

So far, we have made several simplifying assumptions to create stochastic formulations that are more applicable to practical image and signal processing applications. However, even with the assumption of stationarity, it is practically impossible to apply such formulations to some very important applications. At this point, we need to make another assumption to narrow down our focus further to obtain a more practically useful model of stochastic processes. The need and motivation for making more simplifying assumptions is as follows. Note that in real applications, the PDF is not known and must to be estimated from the observed data. To obtain an estimate of the PDF or any statistical averages, we need to have access to several recordings of the process and then perform an ensemble averaging over these recordings to estimate the desired averages. However, in almost all applications, especially in bioinformatics applications, very often only one signal reading is available. As such, it is unreasonable to expect the laboratories to collect a few hundred DNA microarray time series for each case so that we conduct our ensemble averaging to calculate PDF.

From the above discussion, it is clear that in many practical applications, one recording is all we have and therefore we need to calculate the averages such as mean and variance from only one recording. Statistically speaking, this is not feasible unless further assumptions are made. The assumption that helps us with such situations allows us to perform the averaging across time for only one recording and to treat these averages as our ensemble averages. The stationary processes in which the averages of any recording of the signal across time equal the ensemble averages are called "ergodic processes." Formally speaking, considering any outcome signal $x(w_i,t)$ recorded for an ergodic process, we have

$$\bar{g}(t) = E\big(g(x(t))\big) = \int_{-\infty}^{-\infty} g(x(w_i,t))dt \qquad (3.14)$$

Similarly, in discrete ergodic processes using only one recording of the stochastic process, $x(w_i,n)$, we can use the following relation to calculate all ensemble averages through averaging in time:

$$\bar{g}(n) = E\big(g(x(w,n))\big) = \sum_{n=-\infty}^{+\infty} g(x(w_i,n)) \tag{3.15}$$

For ergodic processes, all averages such as mean and variance can be calculated using the above time averaging. Ergodicity is a practically useful assumption made in many biomedical signal and image processing applications, and unless indicated otherwise, all stochastic processes discussed in the book are assumed to be ergodic.

## 3.3  BAYESIAN THEORY

One of the most popular methods in bioinformatics and systems biology is Bayesian theory. The main/original idea of Bayesian theory is to relate the probabilities of two or more phenomenon to their conditional probabilities.

We start the description of Bayesian theory by assuming two random variables $A$ and $B$. These random variables can represent the over/underexpression of two genes in a DNA microarray experiment. Now consider the probability of $A$, probability of $B$, joint probability of $A$ and $B$, conditional probability of $A$ given $B$, and conditional probability of $B$ given $A$ as $P(A)$, $P(B)$, and $P(A|B)$, and $P(B|A)$, respectively. Then Bayesian theory asserts that

$$P(B|A)P(A) = P(A|B)P(B) \tag{3.16}$$

Therefore, we can write

$$P(B|A) = \frac{P(A|B)P(B)}{P(A)} \tag{3.17}$$

This simple Bayesian relationship has tremendous impact on science and engineering, as we will discover later. It has to be mentioned that this relationship can be extended to multiple variables as opposed to only two.

Before moving to the next section, where the application of Bayesian theory in forming probabilistic graphs is described, we need to describe another alternative in writing Bayesian theorem. Since we know that

$$P(B|A)P(A) = P(A,B) \tag{3.18}$$

Then, we can rewrite Equation (3.17) as:

$$P(B|A) = \frac{P(A,B)}{P(A)} \tag{3.19}$$

This form of describing Bayesian theorem is often more popular in statistics books, and we will use it for some of our future calculations.

### 3.3.1  FROM BINARY NETWORKS TO BAYESIAN NETWORKS

Binary networks are popular tools in graph theory and decision-making systems in which a number of binary (0 or 1) nodes are related to each other via some determinist links. Such networks, currently used in systems biology and bioinformatics, are popular for applications such as estimating the network(s) of interactions among genes/gene products, i.e., these links will identify the functional relationships among the genes. For instance, a negative link from gene A to gene B is interpreted as the tendency of gene A to suppress the expression level of gene B.

However, as the aforementioned description of binary networks suggests, these models are determinist models that are typically driven from DNA microarray data, which are noisy and clearly nondeterministic. This simple observation suggests that instead of using determinist models, stochastic solutions might present better approaches for modeling of gene regulatory networks.

Bayesian networks are graphs in which the relationships among the nodes are described using probability measures calculated using Bayesian theory. More specifically, Bayesian networks are directed graphs in which nodes represent variables, and directional links (arrows) represent conditional dependencies among variables. Nodes can represent any kind of variable, be it is a measured parameter, a latent variable, or even a hypothesis.

The details of Bayesian networks will be given in Chapter 8.

## 3.4 TEST OF HYPOTHESIS

Any statement about a population can be supported or denied using samples of data, and a hypothesis test is a method to assess these statements and make statistical decisions. Hypothesis testing can therefore be used in practical decision-making applications. For example, surgeons may need to decide whether to operate on a patient or whether to treat a patient with a particular drug. A hypothesis test could be used to generate advice on the best course of action and support all suggested decisions with statistical analysis.

The testing process consists of four steps. First, the null hypothesis $H_0$ and the alternative hypothesis $H_a$ are formed. A typical null hypothesis is the observations that are the result of pure chance, and the alternative hypothesis is its complement. A statistical test is then chosen, and the $p$ value for the hypothesis is calculated. Small $p$ values suggest that the null hypothesis is unlikely to be true. Finally, the $p$ value is compared with an acceptable significance value, called an $\alpha$ value.

When a hypothesis test is performed using a sample rather than the entire population, there is a risk that an incorrect decision will be made. In this situation, hypothesis testing is a more general way to support a decision.

Two types of error can be made in statistical testing. A type I error, or false positive, occurs when the null hypothesis is rejected when it is actually true. The probability of a type I error is the level of significance of the test. A type II error, or false negative, occurs when a false null hypothesis is not rejected (which should be rejected). Type I errors are usually considered more serious than type II errors and are therefore more important to avoid. This can be done by decreasing the level of significance; however, this in turn increases the probability of a type II error. Choosing a level of significance involves carefully balancing the two types of errors. Increasing the sample size will decrease both error types, but involves extra overheads.

The $p$ value is defined as the smallest significance level at which the null hypothesis is rejected. The advantage of using the $p$ value is that we can see immediately whether our test statistic is in the rejection region. For example, if $p < 0.005$, the chance of rejecting the null hypothesis is below the 5% significance level. Effectively, the $p$ value quantifies evidence in the data against the null hypothesis.

One popular test performed in bioinformatics and systems biology is $t$ test, as described below.

### 3.4.1  $t$ TEST

Suppose that there are two data sets representing a certain characteristic factor. The simplest method to verify if the two groups can be distinguished is to use a $t$ test. A $t$ test, sometimes referred to as Student's $t$ test, measures the similarity of two groups; specifically, whether the means of two groups are statistically different. The basic idea is to compare the sample and population means, with an adjustment for the number of cases in the sample and the standard deviation of the mean. The formula for a $t$ test is

$$t = \frac{x - \mu_1}{\frac{s_1}{\sqrt{n_1}}} \tag{3.20}$$

where $x$ is the mean of a random samples of size $n_1$, $\mu_1$ is the mean of a normal distribution, and $s_1$ is the variance. The computed $t$ value is compared with values in a significance table, depending on the level of significance. In most statistical testing, the $\alpha$ level of 5% is used ($\alpha = 0.05$). A $t$ test should be typically performed with a sample size of at least 30. It is also assumed that the sample is normally distributed.

The two types of $t$ tests are one- and two-sample tests. The single-sample $t$ test is used to compare a single sample with a population with a known mean but unknown variance. For example, we may wish to test the accuracy of an odometer in reporting vehicle mileage. As part of the experiment, we observe that 10 cars travel 10 miles. Now the question is, using the 0.01 level of significance, can the odometer be trusted? First, the null and alternative hypotheses are formed.

$$H_0 : \mu = 10$$
$$H_a : \mu \neq 10 \tag{3.21}$$

Having selected the significance level ($\alpha = 0.1$), next, the degrees of freedom ($df$) calculated must be found. In this case, $df = n - 1 = 10-1 = 9$, where $n$ is the sample size. These two values are used to look up the $t$ distribution value, which is compared with the $t$ test value calculated using Equation (3.21). The result is used to make a decision about the accuracy of the odometer.

A two-sample test can be used to compare two categories of some categorical variable, for example, comparisons made on gender or two populations receiving different treatments in context of an experiment. As an example, we may want to analyze the effects of a drug on patients receiving the drug versus patients not receiving the drug. A two-sample $t$ test tests whether the means of the two populations differ; in other words, whether the use of the drug is associated with significant improvement in a patient's condition. The hypotheses are

$$H_0 : \mu_1 = \mu_2$$

$$H_a : \mu_1 \neq \mu_2 \tag{3.22}$$

The process is identical to the one-sample $t$ test, except that the test statistic is calculated using the formula

$$t = \frac{(x-y)-(\mu_1 - \mu_2)}{\sqrt{\dfrac{(n_1-1)s_1^2 + (n_2-1)s_2^2}{n_1+n_2-2}} \sqrt{\dfrac{1}{n_1} + \dfrac{1}{n_2}}} \tag{3.23}$$

where $x$ and $y$ are the means of two random samples of size $n_1$ and $n_2$, $\mu_1$ and $\mu_2$ are normal distribution means, and $s_1$ and $s_2$ are the variances. The two-sample $t$ test is commonly used in medicine and social sciences to evaluate the effects of interventions and/or treatments.

## 3.5 EXPECTATION MAXIMIZATION METHOD

Expectation maximization (EM) is a general iterative algorithm for finding maximum likelihood (ML) estimates of model parameters in the presence of latent variables. Latent variables are those that are instead inferred from other directly observed variables. The EM algorithm consists of two separate steps: expectation and maximization. The expectation step computes the expected likelihood by treating the latent variables as if they were directly measured. The maximization step maximizes this computed likelihood to generate new ML estimates of the model parameters. These parameters are then used as input to a second expectation step, and the algorithm repeats until convergence.

Formally, the task of EM is to find $\theta$ to maximize $\log f(x|\theta)$ with $x$, where $x$ represents the hidden latent variables. This is done by maximizing the likelihood of the observed measurements $U$, i.e.,

$$\theta = \arg\max_{\theta} \sum p(\theta, x|U)$$

The expectation must be computed over all values of the unobserved variables. Since EM can be guaranteed to reach a local maximum, a global maximum can theoretically be reached, based on the initial values. However, since multiple local maximums are possible, identifying the global maximum can be a challenge. One solution is to try a variety of initial values and choose the highest likelihood value.

EM is typically used for data clustering in machine learning, pattern recognition, and computer vision but is also applied in econometric, clinical, and sociological studies where unknown factors affect outcomes. Because of its generality and guaranteed convergence, the EM algorithm is suitable for many estimation problems.

This method will be further described in Chapter 7.

## 3.6 MAXIMUM LIKELIHOOD THEORY

We start describing the main concepts of ML theory with a concept borrowed from information theory.

### 3.6.1 Kullback-Leibler Distance

The Kullback-Leibler (K-L) distance quantifies the proximity of two probability densities $p$ and $w$; or, in the case of modeling, how close a model distribution $q$ is to the true probability distribution $p$. If $p$ and $q$ are distributions of a discrete random variable, the K-L distance is the expectation of $\log(p/q)$.

$$D_{KL}(p,q) = E\left[\log\left(\frac{p}{q}\right)\right] = \sum_i p_i \log\frac{p_i}{q_i} \qquad (3.24)$$

Or, for continuous variables,

$$D_{KL}(p,q) = \int_{-\infty}^{\infty} p(x)\log\frac{p(x)}{q(x)}dx \qquad (3.25)$$

The K-L distance is nonnegative and is zero only when $p$ and $q$ are equal. However, as it does not satisfy the symmetry and triangle inequality requirements, it is not truly a distance. It is therefore often referred to as the K-L divergence, or as relative entropy. Relative entropy measures the inefficiency of assuming model $q$ when the true distribution is $p$. This is based around the theory of coding and data compression. The most efficient coding is achieved when the true distribution $p$ is known; the resulting code uses an average of $H(p)$ bits to represent a given random variable. Any other distribution will result in less efficient coding, and the amount of extra information to be transmitted is greater or equal to the K-L distance between the two distributions. So, when using a model distribution $q$ instead of the true distribution $p$, encoding the variable requires $H(p) + D(p,q)$ bits.

However, this form of the K-L distance is only useful if the true distribution $p$ is known. In practice, this is rarely the case. By rewriting the equation, the K-L distance can be estimated using just the raw data.

$$D(p,q) = \sum_{i=1}^{m} p_i \log\frac{p_i}{q_i} = \sum_{i=1}^{m} p_i \log p_i - \sum_{i=1}^{m} p_i \log q_i \qquad (3.26)$$

The first term depends only on $p$ and can be treated as a constant unknown value. Therefore, maximizing the second term minimizes the K-L distance. If $p$ is replaced with the uniform distribution, then the right hand side is simplified to $\log q_i$. However, other distributions can be used, depending on the application.

### 3.6.2 Log-Likelihood Function

The K-L distance measures the difference between two different probability distributions. However, when constructing a model, another option is to use the same distribution with different parameters. An accurate model can be obtained by maximizing the model's likelihood. Informally, a likelihood function estimates how likely a particular population is to produce an observed sample. Let $x \to f(x|\theta)$ be a parameterized family of PDFs, where $\theta$ is the parameter. The likelihood function of the parameterized model is

$$L(\theta|X) = f(X|\theta) \tag{3.27}$$

This has the same form as a PDF, except the parameters are emphasized rather than the outcomes. Therefore, while a PDF contains unknown outcomes and known parameters, we can use a likelihood function to estimate parameters given a known set of outcomes. However, the likelihood function cannot be used to calculate the probability that a set of parameters is correct.

The value of $T$ that maximizes $L$ is the ML estimate of $T$. In practice, the ML estimate is often calculated as the maximum of the log-likelihood function, the natural log of $L$. Because many density functions are exponential in nature, this simplifies the calculation.

### 3.6.3 AKAIKE'S INFORMATION CRITERION

Successful model selection requires a well-identified definition of the "best" model, and a prominent approach is Akaike's information criterion (AIC) (Akaike, 1974). AIC is grounded in information theory and formalizes the relationship between likelihood theory and the K-L distance measurement. It numerically evaluates the goodness of fit of a model by measuring the information lost between the true model and the approximation under consideration. The "best" model in a selection is therefore the one with the lowest AIC value, calculated as

$$AIC = -2\log L(\hat{\theta}) + 2k \tag{3.28}$$

where $\log L(\hat{\theta})$ is the maximized log-likelihood function, and $k$ is the number of parameters in the model. Adding more parameters increases the likelihood of the data but can also lead to overfitting, particularly when the number of parameters is large relative to the sample size. To avoid this, a second-order variant of AIC is used:

$$AIC_c = AIC + \frac{2k(k+1)}{n-k-1} \tag{3.29}$$

where $n$ is the sample size and $k$ is again the number of parameters. If the sample size is large relative to the number of parameters, then second-order AIC does not affect the result.

### 3.6.4 BAYESIAN INFORMATION CRITERION

Another well-known model selection method in ML theory is Bayesian information criterion (BIC). The BIC value for a model is given by

$$BIC = -2\log(L) + k\log(n)$$

where $\log(L)$ is the maximized log-likelihood function for the model, $n$ is the sample size, and $k$ is the number of model parameters. BIC penalizes free parameters more strongly than AIC and thus discourages overly complex models.

Unlike AIC, BIC is not based in information theory but is derived from Bayesian concepts; the a priori probability distribution function is used in deriving the rule. BIC is preferred over AIC when the complexity of the actual model does not increase with the size of the data set.

## 3.7  SYSTEM IDENTIFICATION THEORY

It is often the case that an unknown system needs to be modeled using only a set of observations made about the system. These observations are typically formed of inputs (stimuli) fed to the systems and outputs (responses) received from the "black box" system. The theory of system identification is developed to form mathematical models of systems using input-output data that allow accurate models of the unknown systems. These systems are often assumed to be dynamic and have responses that evolve in time. The models used in the system identification theory can be either linear or nonlinear. In this chapter, we focus on one of the most popular linear models and describe the mathematical structure of this system of equations.

### 3.7.1  AUTOREGRESSIVE EXOGENOUS MODEL

One of the most popular models in the system identification theory is the *autoregressive exogenous* (ARX) model. This model relates the future values of the output variables, $y_i$'s, to the values of output variables in the past time(s) as well as the input variables, $x_i$'s. The model also considers the uncertainty inherited in the model by considering a noise factor ($e$) in the equations. Assuming that the system is linear, in its most general form, the mathematical formulation of this dynamic model can be described by the following linear ARX expression:

$$y_i(t) = -a_{i,1,1}y_1(t-1) - ... - a_{i,1,n_a}y_1(t-n_a)$$

$$-a_{i,2,1}y_2(t-1) - ... - a_{i,2,n_a}y_2(t-n_a)$$

$$-...$$

$$-a_{i,p,1}y_p(t-1) - ... - a_{i,p,n_a}y_p(t-n_a)$$

$$+b_{i,1,0}x_1(t) + b_{i,1,1}x_1(t-1) + ... + b_{i,1,n_b}x_1(t-n_b) \qquad (3.30)$$

$$+b_{i,2,0}x_2(t) + b_{i,2,1}x_2(t-1) + ... + b_{i,2,n_b}x_2(t-n_b)$$

$$+...$$

$$+b_{i,q,0}x_q(t-1) + b_{i,q,1}x_q(t-1) + ... + b_{i,q,n_b}x_q(t-n_b)$$

$$+e(t)$$

where $y_i(t)$ is the value of output $i$ ($i =1, 2,..., p$) at time $t$, $x_i(t)$ is the value of input $i$ ($i =1, 2,..., q$) at time $t$, $n_a$ is the degree of the system with respect to the variations in the output, $n_b$ is the degree of the system with respect to the input, coefficient $a_{i,j,k}$ is the parameter relating the output $j$ at time $t - k$ to the output $i$ at time $t$, coefficient $b_{i,j,k}$ is the parameter relating the output $i$ at time $t$ to the input $i$ at time $t - k$, and $e(t)$ is the noise factor. As can be seen, the model considers the outputs in the past time samples to predict the output value at the present sample. Such ARX models constitute the core of system identification theory and are used in many field of research such as systems biology, modeling, and control.

For instance, in a one-dimensional ARX, where only the time activities of one output under the influence of only one input is modeled, the output at time $t$, $y(t)$, is related to the output in the previous times and the input $x(t)$. For such a case, with some simplifications on the notations and the indices, Equation (3.1) is reduced to

$$y(t)+a_1 y(t-1)+...+a_{n_a} y(t-n_a)$$
$$=b_0 x(t)+b_1 x(t-1)+...+b_{n_b} x(t-n_b)+e(t)$$

(3.31)

Such a model allows predicting the value of $y(t)$ based on the previous values as

$$\hat{y}(t|a_1,a_2,...,a_{na},b_1,b_2,...,b_{nb})$$
$$=-a_1 y(t-1)-...-a_{na} y(t-n_a)+b_0 x(t)+b_1 x(t-1)+...+b_{nb} x(t-n_b)$$

(3.32)

where $\hat{y}(t|a_1,a_2,...,a_{na},b_1,b_2,...,b_{nb})$ is the best estimation of the output $y$ at time $t$ given a set of parameters (coefficients of ARX model).

Hereafter, we may use output and "state" interchangeably, knowing that in the training phase, the model is trained only with the observed inputs/outputs. Next, the model parameters must be estimated using the time-series data obtained from experiments and observations. Back to the general case of multidimensional case, we define $\theta$ as the matrix of parameters, i.e.,

$$\theta =$$

$$\begin{bmatrix} a_{1,1,1} & a_{1,1,2} & \cdots & a_{1,1,n_a} & a_{1,2,1} & a_{1,2,2} & \cdots & a_{1,2,n_a} & \cdots & a_{1,p,1} & a_{1,p,2} & \cdots & a_{1,p,n_a} & b_{1,1,0} & b_{1,1,1} & \cdots & b_{1,1,n_b} & \cdots & b_{1,q,0} & b_{1,q,1} & \cdots & b_{1,q,n_b} \\ a_{2,1,1} & a_{2,1,2} & \cdots & a_{2,1,n_a} & a_{2,2,1} & a_{2,2,2} & \cdots & a_{2,2,n_a} & \cdots & a_{2,p,1} & a_{2,p,2} & \cdots & a_{2,p,n_a} & b_{2,1,0} & b_{2,1,1} & \cdots & b_{2,1,n_b} & \cdots & b_{2,q,0} & b_{2,q,1} & \cdots & b_{2,q,n_b} \\ \vdots & \vdots & \ddots & \vdots & \vdots & \vdots & \ddots & \vdots & \cdots & \vdots & \vdots & \ddots & \vdots & \vdots & \vdots & \ddots & \vdots & \cdots & \vdots & \vdots & \ddots & \vdots \\ a_{p,1,1} & a_{p,1,2} & \cdots & a_{p,1,n_a} & a_{p,2,1} & a_{p,2,2} & \cdots & a_{p,2,n_a} & \cdots & a_{p,p,1} & a_{p,p,2} & \cdots & a_{p,p,n_a} & b_{p,1,0} & b_{p,1,1} & \cdots & b_{p,1,n_b} & \cdots & b_{p,q,0} & b_{p,q,1} & \cdots & b_{p,q,n_b} \end{bmatrix}^T$$

(3.33)

Note that to train the model, we need to use the observed time-course data to optimize the values of the parameters. Having formed the matrix of parameters, $\theta$, next we define the matrix of observations, $\varphi(t)$, based on the values of the states and the inputs for $N$ time steps as follows:

$$\varphi(t)=\begin{bmatrix} -y_1(t-1) & -y_1(t-2) & \cdots & -y_1(t-1-N) \\ -y_1(t-2) & -y_1(t-3) & \cdots & -y_1(t-2-N) \\ \vdots & \vdots & \ddots & \vdots \\ -y_1(t-n_a) & -y_1(t-n_a-1) & \cdots & -y_1(t-n_a-N) \\ -y_2(t-1) & -y_2(t-2) & \cdots & -y_2(t-1-N) \\ -y_2(t-2) & -y_2(t-3) & \cdots & -y_2(t-2-N) \\ \vdots & \vdots & \ddots & \vdots \\ -y_2(t-n_a) & -y_2(t-n_a-1) & \cdots & -y_2(t-n_a-N) \\ \vdots & & & \vdots \\ -y_p(t-1) & -y_p(t-2) & \cdots & -y_p(t-1-N) \\ -y_p(t-2) & -y_p(t-3) & \cdots & -y_p(t-2-N) \\ \vdots & \vdots & \ddots & \vdots \\ y_p(t-n_a) & y_p(t-n_a-1) & \cdots & x_1(t-n_a-N) \\ x_1(t) & x_1(t-1) & \cdots & x_1(t-N) \\ x_1(t-1) & x_1(t-2) & \cdots & x_1(t-1-N) \\ \vdots & \vdots & \ddots & \vdots \\ x_1(t-n_b) & x_1(t-n_b-1) & \cdots & x_1(t-n_b-N) \\ \vdots & \vdots & \vdots & \vdots \\ x_q(t) & x_q(t-1) & \cdots & x_q(t-N) \\ x_q(t-1) & x_q(t-2) & \cdots & x_q(t-1-N) \\ \vdots & \vdots & \ddots & \vdots \\ x_q(t-n_b) & x_q(t-n_b-1) & \cdots & x_q(t-n_b-N) \end{bmatrix} \qquad (3.34)$$

The vector of predicted outputs $\hat{Y}(t\,|\,\theta)$ given the set of parameters $\theta$ and observations $\varphi(t)$ is calculated as

$$\hat{Y}(t\,|\,\theta)=\begin{bmatrix} \hat{y}_1(t) & \hat{y}_2(t) & \cdots & \hat{y}_p(t) \\ \hat{y}_1(t-1) & \hat{y}_2(t-1) & \cdots & \hat{y}_p(t-1) \\ \vdots & \vdots & \ddots & \vdots \\ \hat{y}_1(t-N) & \hat{y}_2(t-N) & \cdots & \hat{y}_p(t-N) \end{bmatrix}=\varphi^T(t)\theta \qquad (3.35)$$

To form a criterion for parameter optimization, for each time step, the prediction error function, $\varepsilon(t,\theta)$, is defined as

$$\varepsilon(t,\theta)=Y(t)-\hat{Y}(t\,|\,\theta) \qquad (3.36)$$

where

$$
Y(t|\theta)=
\begin{bmatrix}
y_1(t) & y_2(t) & \cdots & y_p(t) \\
y_1(t-1) & y_2(t-1) & \cdots & y_p(t-1) \\
\vdots & \vdots & \ddots & \vdots \\
y_1(t-N) & y_2(t-N) & \cdots & y_p(t-N)
\end{bmatrix}
\tag{3.37}
$$

Next, we calculate $V_N(\theta)$, which is nothing but the total estimation error between the predicted outputs and the true outputs averaged over all $N$ time points $t = 1,\ldots, N$ as follows:

$$
V_N(\varepsilon)=\frac{1}{N}\sum_{t=1}^{N}\frac{1}{2}[Y(t)-\varphi^T(t)\theta]^T\Lambda^{-1}[Y(t)-\varphi^T(t)\theta]
\tag{3.38}
$$

where $\Lambda$ is a symmetric positive semidefinite $p \times p$ matrix that weighs the relative importance of the components of $\varepsilon$. As can be seen, $V_N(\theta)$ is treated as the overall cost function involved in choosing any set of parameter $\theta$. Then, the parameter optimization problem is reduced to finding a least square sense optimal parameter vector $\hat{\theta}_N^{LS}$ that minimizes the above cost function, i.e.,

$$
\hat{\theta}_N^{LS}=\arg\ \min_{\theta} V_N(\hat{\theta}_N,Z^N)
\tag{3.39}
$$

This leads to the following standard equation least square optimization to find the model parameters in multidimensional cases:

$$
\hat{\theta}_N^{LS}=\left[\frac{1}{N}\sum_{t=1}^{N}\varphi(t)\Lambda^{-1}\varphi^T(t)\right]^{-1}\frac{1}{N}\sum_{t=1}^{N}\varphi(t)\Lambda^{-1}Y(t)
\tag{3.40}
$$

The set of parameters chosen as above will provide the optimal model of the biological process under study and the pathways involved in it.

### 3.7.2 LEAST SQUARE ESTIMATION

In many problems in systems biology as well as in many others fields of study, such as the ARX modeling discussed above, one would need to optimize a set of parameters based on a number of observations. We formulate a problem in the matrix form to have a more generalized notation for the parameters. Let us assume that all parameters to be optimized/searched for are arranged in a vector $P$, where $P$ is $n \times 1$. In other words,

$$
P=[p_1\ p_2\ p_3\cdots p_n]
\tag{3.41}
$$

Next, assume that every observation that we have about these parameters is in the form of the following linear equation:

$$
c_{i1}p_1+c_{i2}p_2+\cdots+c_{in}p_n=g_i
\tag{3.42}
$$

Assuming that we have $m$ such observations, we will have a set of $m$ equations with $n$ unknown $p_i$'s. We can write this set of scalar equations in a matrix form:

$$CP = G \tag{3.43}$$

where $C$ is $m \times n$ and $G$ is $m \times 1$. If $m = n$, the solution for the above problem, if it exists, can be found rather easily. All one needs to do is to multiply both sides of the equation with $C^{-1}$, if it exists, and that gives $P$, which is the set of parameters we are looking for. If $m < n$, then one needs to make more observations before solving the problems because the number of equations is less than the number of unknowns. Finally, if $m > n$, the number of equations is more than the number of unknowns. In such a situation, one cannot find a set of solution that completely satisfy all equations (constrains) at the same time. However, it is almost always desirable to find a set of solution whose error in satisfying all of these equations is minimal. Specifically, if you define the error in satisfying these equations in squared distance sense, it would be advantageous to find a solution that minimizes the average (mean) of these errors. Such a solution is called least mean square error, or least square error.

Next, we explore the solution for $m > n$ case, which is the dominant case in many practical application. Starting from Equation (3.44), since we cannot define an inverse matrix for a nonsquare matrix like $C$, we add an extra intermediate step and multiply the two sides by $C^T$:

$$C^T CP = C^T G \tag{3.44}$$

Now, note that the matrix $C^T C$ is a square matrix and can have an inverse. Assuming that this inverse exists, we multiply the two sides of Equation (3.45) by this inverse:

$$(C^T C)^{-1}(C^T C)P = (C^T C)^{-1} C^T G \tag{3.45}$$

Simplifying the left side of the equation:

$$P = (C^T C)^{-1} C^T G \tag{3.46}$$

This gives the least square solution to the problem we were trying to solve, i.e., finding the best set of parameters to match the observations.

## 3.8  SUMMARY

In this chapter, a number of computational techniques were reviewed. These techniques included the basic methods in probability theory. The theory of stochastic processes was briefly discussed in this chapter. The basic principles and applications of the system identification theory were also reviewed.

## 3.9  PROBLEMS

1. A probability distribution function is given as follows: $p(0) = 1/4$, $p(1) = 1/3$, $p(2) = 1/4$, $p(3) = 1/6$. For this PDF, calculate the
   a.  Mean
   b.  Variance
2. In this chapter, we described ARX model in which an exogenous input was included in the model/system. Consider the case in which there is no exogenous input, e.g., there is no drug or environmental factor. In such studies, the normal progression of the system is studied without any intervention. The model used to describe such a system is the autoregression model, in which every part of the model is exactly the same as ARX (described above) except that the exogenous input is entirely removed from the formulation. Using the formulation of ARX given above, write the entire formulation for the autoregression model.
3. Focus on the definition of stationary random processes given above and
   a.  Describe three biological systems that are stationary
   b.  Describe three biological systems that are not stationary
4. Focus on the definition of ergodic random processes given above and
   a.  Describe three biological systems that are ergodic
   b.  Describe three biological systems that are not ergodic

# 4 Computational Structural Biology
## *Protein Structure Prediction*

## 4.1 INTRODUCTION AND OVERVIEW

The function of a protein is largely a consequence of its structure. Hence, as our understanding of a protein's structure progresses from its primary structure (its sequence of amino acids) to its secondary structure (its subordinate, local 3-D structures such as alpha helixes and beta strands), to its tertiary structure (the full, 3-D folded conformation), to its quaternary structure (functional complexes with other proteins), our ability to understand the protein's function likewise improves. Were an accurate method of predicting a protein's tertiary structure directly from its primary structure—a so-called de novo prediction of protein structure—to exist, it would unlock a wide array of extraordinarily valuable applications, from drug design to medical diagnostics. Some of the complicating factors that confound de novo prediction of protein structure include

1. Approaches that rely on molecular dynamics (MD) are simplifications that still require massive amounts of processing power.
2. Homology searches are inherently fuzzy, in that key residue substitutions or changes in order may still allow for good sequence matches, but may have substantial impact in the fully folded conformation of the protein.
3. Machine-learning and statistical methods require large volumes of consistent data, but each primary structure may result in multiple tertiary structures. Proteins can switch conformations thousands of times a second, so even database structures are best termed *ensemble structures*, meaning that they are the average or consensus structure resulting from analysis. Furthermore, the distribution of these conformations may be significantly influenced by environmental factors such as activating proteins, solvent interactions, or ligands.

There is a significant amount of active research being conducted toward understanding of how to predict protein structure from a number of different perspectives.

The remainder of this chapter will examine some broad classes of protein structure prediction methods and briefly present some of the data resources that support these methods.

## 4.2   PROTEIN STRUCTURE PREDICTION METHODS

Predicting the structure of a folded protein from nothing more than its ordered list of amino acids is a very difficult task. There are many different tools that try to predict protein structure, and they can be classified according to a variety of dimensions, but a simple measure that helps to organize them for this presentation is the level of background knowledge that they incorporate. In this light, we will discuss three classes of methods: ab initio methods, comparative methods, and hybrid/informed.

### 4.2.1   AB INITIO METHODS: MOLECULAR DYNAMICS

At one extreme of the continuum of protein structure prediction methods is the set of MD methods. These are the approaches that are grounded in nothing more than assumptions about how individual atoms and small molecules interact (hence the term "ab initio"). Clearly, this is about as low-level, information-less approach as is possible. Just as clearly, an approach that begins at this low level requires a significant amount of computing power; so much power, in fact, that until recently, these approaches have been relegated to working on very small problems. With the advent of exponentially more powerful machines and faster networking, however, more interesting problems are becoming computationally tractable.

Examples of the MD approach to predicting protein structure include

IBM's Blue Gene/L project: http://www.research.ibm.com/journal/sj/402/allen.html. Taking advantage of IBM's investment in massively parallel computing, Blue Gene/L is a software project married to a very large hardware installation, relying upon the supercomputing capabilities to support the number of calculations required for MD predictions. Because of the implicit hardware availability, the Blue Gene/L project also includes solvent interactions.

Folding@Home (FAH): http://folding.stanford.edu/, powered by GROMACS (http://www.gromacs.org/). Because it is an MD approach, FAH also requires a huge amount of processing power; unlike Blue Gene/L, however, FAH relies upon a federation of small computers acting in concert. While this adds issues such as the reliability of these machines (and unpredictable availability of any given node in the network), it has the advantage of being a much less expensive solution up front.

### 4.2.2   COMPARATIVE METHODS

In contrast to ab initio methods of protein structure prediction, comparative methods do not attempt to begin from first principles but instead take advantage of work that other researchers have done (and collected) on the structure of other proteins. Comparing the new (target) protein with existing databases of proteins and their structures can suggest folded conformations that may be more likely.

There are two general classes of comparative methods: homology-based inferences and threading models. Each is treated here in a separate section.

#### 4.2.2.1 Homology-Based Methods

Homology-based methods are founded in the precept that a good agreement between two proteins' sequences ought to entail similarity in their folded conformation. The key to the success of these methods, of course, is a very reliable way of assessing the quality of match between the two proteins' primary structure. Chapter 5 deals with sequence analysis issues directly. The other prerequisite is having reliable stores of well-established protein structures; fortunately, these exist and are discussed in a later section within this chapter.

#### 4.2.2.2 Threading Methods

Threading is an approach in which the target protein's sequence of amino acids is mapped onto the sequence of a second protein whose folded conformation is known. The shape into which the target protein has been threaded is assigned an energy level based on the bond angles that result from the threading. Repeating this process against a number of known proteins allows the researcher to identify the match whose threading results in the lowest energy mapping.

There are multiple levels at which threading can occur: It can occur at the full 3-D structural level or it can be simplified to apply merely to the fold level. The former is closer to what most protein structure prediction methods desire, but the latter is much less computationally intensive and may well qualify as "good enough" for many purposes.

### 4.2.3 HYBRID/INFORMED METHODS

Between the ab initio approaches that begin by assuming no information about the final conformation and the homology-based approaches that merely reuse the conformation of a previously analyzed protein are the hybrid or informed methods. These are the methods that use databases of existing protein structure to inform their search for building a prediction of the structure of the target protein. These approaches typically represent a statistical or data mining approach.

We will briefly introduce a few different methods here, including

Artificial neural networks (ANNs)
Hidden Markov models (HMMs)
Support vector machines (SVMs)
Clustering
Multimethod approaches

#### 4.2.3.1 Artificial Neural Networks

ANNs have been around since the 1940s and are a machine-learning method that builds structural summaries of data and patterns. Constructed from individual neurons that each accept inputs and perform some simple computation to produce an output, the ANN itself is not a particularly complicated design. The construction of the network itself is tricky: deciding how many input neurons to have, establishing the encoding scheme to use for the inputs, selecting the number of hidden layers to put in the network, and describing the outputs appropriately. Once they are built, ANNs use and rely upon a relatively large amount of data for training; in this case, training refers to the adjustment of weights between connected neurons, so that the input data are best mapped to the output results.

ANNs offer the ability to solve nonlinear problems, but the network structuring can be difficult; even good researchers can get tripped up on the encoding, and many networks require more data to be trained reliably than what users assume. Fortunately, convention can solve the first two issues, and the rapidly expanding protein structure databases—addressed later in this chapter—help to solve the data paucity issue.

Most ANNs used to predict protein structure use a sliding-window approach in which every sequence of $N$-contiguous residues are presented to the network as a single input instance, and the output of the network is graded against the accuracy of the prediction for the residue in the middle of the input sample. That is, if the window size is 17 residues, then the goal of the network is to predict the secondary structure correctly for the ninth residue in the input sequence: What secondary structure does it have? If the network predicts correctly, the training algorithm reinforces the weights that produced that correct result; if the network prediction is wrong, then the weights are adjusted to reinforce what should have been the correct answer. In this way, the network ought to improve its predictions over the training epochs.

Rost and Sander's PHD is a neural network that uses multiple alignment data to predict secondary structures within a protein. In their 1993 paper, Rost and Sander report a 69.7% predictive accuracy, using two networks. The first—which predicts the secondary structure for a residue—is a network having 13 residue positions (819 nodes) represented in the input layer, 6 nodes in the hidden layer, and 3 nodes in the output layer (one for alpha helix, one for the beta strand, and one for the loop structure).

### 4.2.3.2  Hidden Markov Models

HMMs are another machine-learning method that can be used to summarize patterns in data, specifically modeling the probabilistic transition among different states as well as states to observable properties. Unlike regular Markov models, in which the model parameters are visible, an HMM involves either states or transition probabilities that are not directly observable. Depending on the nature of the data available, and the problem to be solved, there are different algorithms that can train the model against the data.

For proteins, there are many possible states that could be represented, but these are most likely to relate to properties of the primary sequence of amino acids. Encoding, as with ANNs, is an issue: Is a state named for its residue, or does it relate to some other property (such as hydrophobicity)? This influences the transitions between states that are to be part of the model. Typically, the outputs (observations) are related to the secondary structures we wish to predict.

The sequence alignment and modeling system* uses HMM tools to encapsulate patterns of data that characterize protein families. The model states are constructed to represent sequence alignments among family members, allowing the HMM to

---

* http://www.soe.ucsc.edu/research/compbio/sam.html

return predicted alignments for a target protein. Sequence alignment and modeling also has an iterative model that interacts with protein databases to find best matches for a target sequence.

### 4.2.3.3 Support Vector Machines

SVMs are another machine-learning method—in fact, a subset of ANNs—that are particularly useful when you are trying to separate data instances into two distinct categories. They do this by the network to maximize the distance between members of the two groups. Each of the neurons uses a kernel function.

As with other ANNs, SVMs present challenges in the form of data encoding. Fortunately, SVMs have a slightly more standardized network structure, which removes some of the ambiguity in design, but there are additional parameters that must be selected/trained before one can expect reasonable performance: In the case of radial-basis functions as the kernel function, for example, C and gamma must be tuned for the SVM to product meaningful results. Also, because SVMs are often used for binary classification, predicting secondary structure (three-valued: alpha, beta, and loop) requires special consideration, often by building cascading networks of SVMs in which the top SVM decides whether the sequence is secondary structure element (SSE) 1 or not, and the second (corresponding to the *not* case) distinguishes between SSE 2 and SSE 3.

Zimmerman and Hansmann's (2008) LOCUSTRA is just such a cascading SVM model used for predicting the secondary structure for a windowed sequence of amino acids.

### 4.2.3.4 Clustering Methods

Clustering methods are based on the principle that single-match homologies may be overspecific and, hence, brittle. Instead, clustering tools tend to build consensus prototypes among the matches. Consider the results of searching for a target sequence in protein structure database: If you review the top three matches, they will likely differ in their allocation of secondary structures, so what the clustering methods attempt to do is to ameliorate these differences in some sort of average or prototype result structure.

MaxCluster* is an example of software that uses protein clustering. Although it is primarily used for sequence alignment, it does incorporate—among others—the 3D-jury† method.

### 4.2.3.5 Blended Methods

Blended methods are simply combinations of other methods. The motivation for doing so is to balance the power and shortcoming of the constituent methods. If a full MD approach is infeasible because of a shortage of computing power, then perhaps a reduced MD method can be applied to extend the results of homology search or threading to refine the quality of the predicted secondary structures, or perhaps, families of proteins can be characterized well by HMMs, but within

---

* http://www.sbg.bio.ic.ac.uk/~maxcluster/download.html
† http://meta.bioinfo.pl/submit_wizard.pl

each family, a custom, cascading SVM might do the best job of predicting secondary structure. When these different methods are combined into a single solution, the resulting hybrid may be able to outperform any of the constituent methods individually.

The Rosetta Commons project* is built around a core suite of products that employ a variety of methods to perform secondary sequence prediction. The ab initio product, for example, uses simulated annealing—another machine-learning method, inspired by metallurgy—to select low-energy combinations of secondary structures suggested by other matching subsequence fragments suggested by other means.

## 4.3   DATA RESOURCES

As was mentioned earlier, many of the prediction methods are founded in the assumption that there are reliable repositories of known structural information for representative proteins. The domain of raw sequence and structure data has been well studied and has benefited from significant consolidation and standardization. A second set of repositories, those dedicated to analysis, annotation, and classification of these structures, is still enjoying a growth phase. Each of these types is described here briefly.

### 4.3.1   PROTEIN SEQUENCE AND RAW STRUCTURE DATABASES

The Protein Data Bank (PDB), dating from 1971, was one of the first repositories for information about the raw 3-D structure of proteins and DNA fragments. In 1998, management responsibility for the PDB transferred from Brookhaven National Laboratory to the Research Collaboratory for Structural Bioinformatics (RCSB). PDB continues to grow rapidly, as do most of the repositories for protein and nucleotide data. This database is accessible via http://www.pdb.org. The PDB is now more significant as a participant in the Worldwide Protein Data Bank, established in 2003, a consortium of similar repositories from across the globe: http://www.wwpdb.org/. The Research Collaboratory for Structural Bioinformatics PDB is still the root of this organization and is the sole gateway for new data that are deposited, although all members agree to synchronize their copies of the database.

A variety of information for each structure is available in PDB, including sequence details, atomic coordinates, crystallization conditions, 3-D structure neighbors, derived geometric data, structure factors, 3-D images, and a variety of links to other resources.

By virtue of having been the first large player in the space, the PDB file formats—ASCII-based and described at http://www.wwpdb.org/docs.html—are the lingua franca of protein data exchange. These formats enjoy wide support throughout the communities of bioinformatics researchers. The proteins within the database also share a labeling convention, the PDB ID.

---

* http://www.rosettacommons.org/software/

## 4.3.2    Protein Structure Classification Databases

In addition to the Worldwide Protein Data Bank, which stores primarily sequence and raw structural data, there are additional databases that store information about how to classify proteins according to their structures. This is advantageous as a means of grouping proteins either for functional or evolutionary comparisons.

The methods used to classify the protein structures in these databases vary from manual examination of structures to fully automatic computer algorithms, to hybrid approaches that combine both human and computer analyses.

The databases briefly presented in this section include

Structural Classification of Proteins (SCOP)
Class, Architecture, Topology, and Homologous Superfamily (CATH)
The Molecular Modeling Database (MMDB)
Families of Structurally Similar Proteins (FSSP)/Distance Matrix Alignment
    (DALI)
Spatial Arrangement of Backbone Fragments (SARF)
InterPro

### 4.3.2.1    The SCOP Database

The SCOP* database is housed at Cambridge and has classified proteins—both those curated manually as well as those that were analyzed using software tools—to show structural and evolutionary relatedness. The hierarchical levels—in decreasing order of strength of relationship—are family, superfamily, and fold.

The "family" level is reserved for proteins that strongly suggest they have a common evolutionary past, both because of their sequence similarity and because of their functional domains. The next strongest level of association is the "superfamily" level, in which proteins are grouped together because there *may* be a shared evolutionary history; generally, there is not enough agreement in either sequence or function to warrant sharing the same family class. Lastly, and most weakly, the "fold" level is reserved for those proteins that share some sequence and structural similarities but without enough support to make any stronger claim on their relationship.

The core principle is that the shared evolutionary history of proteins is directly related to the story of their function. If it were possible to recreate the evolutionary trajectory of all protein variants, then that tree could be annotated with structural and functional adaptations over time. It is not possible to recreate this tree perfectly, of course, so SCOP uses its three-level classification hierarchy as a stand-in method to group proteins according to their commonalities.

### 4.3.2.2    Class, Architecture, Topology, and Homologous Superfamily

The CATH[†] protein structure database at University College London, is similar in overall approach to SCOP, but CATH classifies proteins into four hierarchical levels

* http://scop.mrc-lmb.cam.ac.uk/scop/
† http://www.cathdb.info/index.html

of ascending specificity: class, architecture, topology, and homologous superfamily. These protein classifications are meant to inform researchers searching for similar entries to use as the basis for inferred function; together, their initials form the site name. Membership is determined through a combination of manual and automatic methods: If the innate similarity between two sequences is very high, then the classification is inherited from a known protein to an unknown protein; otherwise, manual processes—relying upon output from automatic analysis methods—kick in to assign an appropriate classification.

The "class" level group proteins are based on the predominant SSEs: the four main groups are those dominated by alpha structures, beta structures, a combination of both, and no abundance of either SSE. This is the lowest level of structural data used in the CATH classification scheme.

The "architecture" level uses descriptions of macrostructures of SSEs to identify similarities among proteins. Proteins that share a "mainly beta two-layer sandwich" description, for example, may share the same architecture. This level of classification uses higher-order data than "class" but more granular data than the "topology" level.

The "topology" level uses slightly more abstract data than the "architecture" level, relying on the general shape of the protein as well as the specific transition between SSEs to group proteins. Because this level is concerned with the way in which the entire protein folds into its final state, this level is also called the "fold family" classification.

The "homologous superfamily" level is for grouping proteins together based on supposed homology (based on cutoff criteria from standard analysis tools). Proteins that share a classification at this level can be further differentiated using their SOLID subclustering, a method based on both sequence identity and overlap from a Needleman-Wunsch analysis.

### 4.3.2.3  Molecular Modeling Database

Proteins of known structure from the PDB have been categorized into structurally related groups in MMDB by the Vector Alignment Search Tool structural alignment program. This program aligns 3-D structures to facilitate the search for similar arrangements of secondary structural elements. MMDB has been incorporated into the ENTREZ sequence and reference database at http://www.ncbi.nlm.nih.gov/Entrez.

Because the entries within the MMDB have had their 3-D structures compared against all other entries, there are well-defined neighborhoods of similarity for each candidate 3-D structure (protein). This allows researchers to review these closest structural neighbors for clues as to function or important motifs.

### 4.3.2.4  Fold Classification Based on Structure-Structure Alignment of Proteins (FSSP/DALI) Database

The FSSP/DALI* database contains structural alignments for proteins in the PDB. The classification and alignments are updated using the DALI structural alignment software.

---

* http://ekhidna.biocenter.helsinki.fi/dali_new/start

Based on the PDB90—in which sequences that are more than 90% similar are eliminated because they provide no additional structural information—the FSSP/ DALI search tool allows you to find a *representative* structure for any PDB entry. This database is updated regularly, performing an exhaustive comparison of sequences against each other in an entirely automatic manner.

DALI was developed by Holm and Sander and published in 1993. This is a matrix-based method that uses distances between amino acids to define patterns of connectivity. These patterns are then used to find matches exhaustively within the database and can be used to score the quality of these matches.

### 4.3.2.5  Spatial Arrangement of Backbone Fragments Database

The SARF* database provides a protein database on the basis of structural similarity, principally relying upon SSEs. The central idea is that, because the overall structure of a protein is closely related to the development of its SSEs, then a good match (relatively low root mean-squared error, RMSD) between the SSEs of two proteins ought to motivate a similarity in structure and, hence, perhaps function.

The SARF Web site provides a similarity-based tree of structures at http://123d. ncifcrf.gov/tree.html.

### 4.3.2.6  InterPro

The European Molecular Biology Laboratory–European Bioinformatics Institute (EMBL-EBI) leads a consortium of protein structure collectors that consolidate their data into InterPro.[†] InterPro acts as a portal, allowing search queries that are federated among the member databases; results are presented together in a single page.

Some of the other consortium member databases include

PROSITE: part of the ExPASy family; http://www.expasy.ch/prosite/
PRINTS: http://www.bioinf.manchester.ac.uk/dbbrowser/PRINTS/
ProDom: http://prodom.prabi.fr/prodom/current/html/home.php
Pfam: part of the Sanger Institute; http://pfam.sanger.ac.uk/
SMART: Simple Modular Architecture Research Tool; http://smart.emblheidel berg.de/
TIGRFAMs: families generated using HMMs; http://www.tigr.org/TIGRFAMs /index.shtml
PIRSF: classification system; http://pir.georgetown.edu/pirwww/dbinfo/pirsf. shtml
SUPERFAMILY: also uses HMMs; http://supfam.cs.bris.ac.uk/SUPERFAMILY/
PANTHER: Protein ANalysis THrough Evolutionary Relationships, maintained by the Evolutionary Systems Biology Group at SRI; http://www.pantherdb.org/
Gene3D: hosted by University College London; http://gene3d.biochem.ucl. ac.uk/Gene3D/

---

* http://123d.ncifcrf.gov/sarf2.html
† http://www.ebi.ac.uk/interpro/

## 4.4  SUMMARY

Protein structure is the key to protein function, so a better understanding of how to quantify, compare, and predict protein structures will contribute directly to a better understanding of how proteins function. There are many different approaches to these problems, many of which have been presented in this chapter. Of note is the contribution made by many different fields (typical of systems biology problems), from biology to mathematics to machine learning and artificial intelligence. All of these methods are enabled by the diligent collection, organization, and curation of accurate biological data in shared, public repositories.

## 4.5  PROBLEMS

1. Why is it that a protein's primary structure entails its higher-order structures?
2. What is the most significant reason that MD approaches do not run as desktop applications?
3. What are some of the challenges of applying ANNs as predictive models of protein structure?
4. What is the difference between a regular Markov model and an HMM?
5. In what way is the name "threading" indicative of the process?
6. Find three protein structure databases *not* mentioned in this chapter; to what audiences are they geared? Do they participate in larger collections? Why?

## REFERENCES

Holm, L., and C. Sander, "Protein Structure Comparison by Alignment of Distance Matrices." *Journal of Molecular Biology*, 233 (September 1, 1993): 123–138.
Rost, B., C. Sander, and R. Schneider, "PHD—An Automatic Mail Sever for Protein Secondary Structure Prediction." *Compute Appl. Biosci.,* 10 (February 1994): 1, 53–60.
Rost, B. and C. Sander, "Improved Prediction of Protein Secondary Structure by the Use of Requence Profiles and Neural Networks." *Proc, Natl. Acad. Sci.,* 90 (1993): 7558–7562.
Zimmermann, Olav, and Ulrich H. E. Hansmann, "LOCUSTRA: Accurate Prediction of Local Protein Structure Using a Two-Layer Support Vector Machine Approach." *Journal of Chemical Information and Modeling*, 48 (2008): 9, 1903–1908.

# 5 Computational Structural Biology
## *Protein Sequence Analysis*

## 5.1 INTRODUCTION

Protein and nucleotide sequences play an important role in bioinformatics. Sequence analysis concerns the various methods used to glean information from these linear strings of amino acids and nucleotides. There are two categories of sequence analysis, one of them focusing on analyzing a single sequence and the other on aligning two or more sequences. Both of these approaches provide useful information to researchers, and both of them are used extensively.

Single-sequence analysis means different things depending on whether the sequence is in protein or in DNA. The preceding chapter, for example, was dedicated to the prediction of protein structure from its sequence of amino acids. For DNA, it does not make any sense to ask how the sequence of nucleotide basepairs affects structure but rather to ask how the sequence of base pairs does (or does not) interact with the proteins in the nucleus to encode and express different genes. Because single-sequence analysis is so specific to the strand being studied, it is not covered in any greater detail here; these domain-specific topics are addressed in other chapters within the text.

In pairwise alignment, one wants to find the similarity between two polypeptide or nucleotide sequences. This sort of analysis is important in evolutionary terms because two homologous sequences that have a common ancestor can be placed in a phylogenetic tree by measuring the similarity between them. These methods can also help researchers find similar sequences in a database or to find a similar region to a given sequence (called the query sequence) in a database. Pairwise sequence alignment may be used to predict sequence function by summarizing known functions of other sequences with which the query string has a high similarity.

But similarity is a quality that is quantitatively measured. Therefore, it is necessary to compose the quantitative measurement carefully. The significance of alignment helps researchers estimate the probability of finding similar sequences by random chance.

Multiple-sequence alignment is an important type of analysis that can explore the similarities and differences in more than two sequences at one time. We discuss this issue in the last section and introduce the various methods of multiple alignment for both nucleotide and polypeptide sequences.

## 5.2  PAIRWISE SEQUENCE MATCHING

In this section, we first want to find a way to express the similarity of two sequences, and then we want to use this capability to find good matches to our target sequence within a larger collection. Similar sequences, if we could identify them quickly and easily, may also share similar structure and functionality, even in different species. A good pairwise sequence matching scheme, then, benefits both database searching as well as evolutionary analysis.

Furthermore, the similar sequences in different species may have been emerged from a common ancestor. One can calculate the number of evolutionary mutations from alignment of two similar sequences and construct phylogenetic trees that show the relationship among a collection of different organisms, all from sequence alignment data.

### 5.2.1  Pairwise Background

Before any discussion, we first describe the concept of similarity of two sequences. Unfortunately, despite homology, similarity is a slippery concept to define. Similarity is rooted in purpose, and our principal purpose is to allow the calculations to be practical. One can calculate the similarity score simply by assigning a merit score to the matching words and penalty score to the mismatches. Generally, matches have higher scores than mismatches because it is matches what we seek.

There may also be two homologous RNA sequences with some interior regions not in common to both strands, perhaps attributable to introns. If one calculates the score and permits gaps—termed INDEL regions, derived from insertion or deletion of residues—between words, it may produce a higher score. In such cases, one should set a gap penalty for preventing unessential splicing sequences.

Setting the gap penalty for each INDEL word has an obvious drawback: It assigns the same penalty for every word within a large gap, which is unrealistic. In biology, a continuous gap (missing two or more contiguous words in a sequence) often refers to the splicing region of RNA sequence or amino acid chain. In these cases, the presence of the gap represents a greater penalty than does the length of the gap, so the first residue of a gap should incur a high penalty (gap open penalty), and each subsequent residue within the continuous gap should have a lower penalty (gap extension penalty).

There are two different methods for pairwise sequence alignment. In global alignment, the best overall alignment between two sequences is sought whether local, high-scoring subsequences are in alignment or not. The other method, which is called local alignment, seeks high-scoring subsequences and aligns them together among sequences.

### 5.2.2  Sequence Matching Methods

There are various methods for pairwise sequence matching that favor local or global alignment. There are many types of software using these methods either to align two sequences or to search a database for finding similar sequences. Using any of these methods has both strengths and weaknesses.

### 5.2.2.1 Dot Plot Visualization Method

Dot plot is one of the oldest and simplest methods of pairwise sequence alignment. To construct a dot plot, one of the sequence's elements is used to define the columns of the matrix and the other sequence's elements define the rows. For each cell, if its row symbol (from the first sequence) and column symbol (from the second sequence) are the same, we fill the cell (place a dot within it); otherwise, the cell remains empty. If the two sequences were identical, then the identity line would be filled with dots and all other cells that had dots would be merely spurious matches.

A dot plot is often a noisy matrix, and we can filter (denoise) it in various ways. One of the simplest and most applicable methods of denoising is the window/stringency method, which specifies a window size and a stringency that refers to the minimum number of matches required within each window. Within every diagonal window of $W$ consecutive cells, at least $S$ (the stringency) must contain dots (matches). For example, if the window size is set to 2 and the stringency is set to 1, a dot remains full if and only if at least one of its two diagonal neighbors is also filled.

Diagonal lines in this matrix represent regions of similarity between the two sequences. The filled cells within a diagonal represent matching characters, and the empty spaces in diagonals refer to the INDEL regions. Thus, a dot plot matrix shows a rough global alignment of two sequences, highlighting (via the diagonals) the regions that show good contiguous matches. If we allow inverted repeats, we can find complementary regions in two sequences that, for example, can be used to determine folding of RNA molecules or for finding two sense/antisense strands of RNA molecules.

The window/stringency numbers frame the desired accuracy of visualization. If the window size is large, the locally aligned subsequences we seek will be longer, and if the difference between window size and stringency is high, it means that we are applying only a very coarse filter. Typically, window/stringency for nucleotide sequences is 15/10 and for polypeptide sequences is 1/1 or 3/2, indicating that, for polypeptide sequences, a narrower sliding window with high accuracy is more appropriate.

Window/stringency values are often established heuristically. For example, one can use substitution values (discussed later in this chapter) or dynamic averages for a more useful dot matrix. In these variants, a window will be preserved only if the number of its hits (dots, matches) is more than a certain average score.

### 5.2.2.2 Dynamic Programming

Dynamic programming is an algorithm, not a heuristic,* which helps to establish the optimum alignment (allocation of matches, mismatches, and gaps) between two arbitrary sequences.

Assume that we have two particular sequences X and Y with lengths $m$ and $n$, respectively. If the objective is to find the best comparison score of these two sequences, one must align each subsequence of the length $m + n$. The number of

---

* The key difference between the two is this: Algorithmic search tends to be exhaustive, whereas heuristic search tends not to be. The trade-off is between time and completeness.

possible alignments grows exponentially. For practical purposes, some other method must be used.

Dynamic programming is a computational method that is able to solve optimization problems by dividing them into independent subproblems. It is a recursive approach that saves its intermediate results in a matrix where they can be used later in the program (trading memory space for execution time). Dynamic programming can be used to solve our alignment problem. In this case, the subproblem is alignment of prefixes of two sequences that are calculated and stored in a matrix. The three steps in dynamic programming are (1) initialization of the matrix, (2) filling the matrix, and (3) traceback from the desired end state.

Dynamic programming guarantees to find the best local or global alignment (providing that a reasonable scoring function is used) because, in this algorithm, all of the possible alignments have been calculated. However, this algorithm is computationally intensive. For example, short sequence alignments by this algorithm may need more than an hour to produce results. To address this problem, many heuristic-based methods that are variants of (or rely upon) dynamic programming have been developed. These methods often are called word methods.

### 5.2.2.3 Word Methods

The word methods—heuristics, rather than algorithms—for pairwise sequence alignment use multiple polypeptides or nucleotides as a single word unit for their alignment instead of a single polypeptide or nucleotide. The two famous programs that use word methods are BLAST and FASTA. These two programs use different types of word methods, but both of the algorithms are fast enough to search for the alignment of a query sequence against an entire database. These methods are commonly used for database searching, whereas dynamic programming alone is generally time-prohibitive.

When searching for matches for a new DNA or protein sequence, one typically does not know whether the expected similarity that he/she is searching for refers to the whole query sequence or just a part of it. Therefore, one can search the database with an option for computing the best local alignments.

FASTA algorithm, which predates BLAST, uses words of two amino acids long for polypeptide alignment ($20^2 = 400$ possible words) or words of six nucleotides long for alignment of nucleotide sequences ($4^6 = 4,096$ possible words). In the first step of the algorithm, FASTA builds the dot plot matrix from the query sequence and the database entry sequence. In the next state, FASTA scores the top 10 alignments (without gaps) that contain the most similar words. It uses dynamic programming-like methods for this local alignment. The scoring scheme is the Blosum 50 substitution matrix, discussed later in this chapter. In the final state, the top 10 sequences with the highest nongapped similarity with the query sequence are merged into the gap alignments to produce the "optimized score."

FASTA also calculates the expectation number, $E$, that represents the number of expected random alignments that have greater alignment scores with query sequence than "optimized alignment." This value estimates the statistical significance of the results.

The BLAST program is another program that uses the word method for pairwise sequence alignment. The word length of polypeptide sequences in BLAST is typically three characters, and the length of nucleotide words is commonly 11 bases. The advantage of BLAST is that the top score sequences (called maximal scoring pairs) are extended to the maximum possible length for a particular alignment score.

There exist excellent Web servers that offer these programs—in particular, at the European Bioinformatics Institute and National Center for Biotechnology Information (NCBI)—where BLAST and FASTA can be used on up-to-date DNA and protein databases.

### 5.2.2.4 Bayesian Methods

The Bayesian method of pairwise sequence alignment is a statistical method that is founded upon conditional probabilities. It can determine the joint probability of two events using their prior probabilities and initial information. We discuss the Bayesian statistics more in the next chapter. By using this method, one can estimate the evolutionary distance between two DNA sequences (a task that resembles pairwise sequence alignment). The Bayesian assumption is that the alignment of two sequences is based on both the prior probabilities of (mis) matches between residues and the probability of gaps. The alignment probability is called posterior probability, and the initial information called prior (or anterior) probability. This allows the Bayesian method to estimate the uncertainty of the alignment by deriving the exact significance measures. This is the key advantage of Bayesian method in comparison with dynamic programming and word methods.

In practice, the Bayesian tools, like Bayes block aligner, perform better than dynamic programming in some cases and worse in the others. The Bayesian algorithm for local sequence alignment is another program that uses the Bayesian method.

### 5.2.3 STATISTICAL SIGNIFICANCE

The statistical significance of a certain pairwise alignment score for two sequences is the probability of observing better scores if the target sequence were aligned with all random sequences. In other words, the significance of an alignment describes the likelihood of its occurring by chance. If we can model the score of alignment with random sequences as a random variable driven by a known distribution function, then we can easily assign significance to each alignment score by calculating the cumulative distribution function.

## 5.3 MULTIPLE SEQUENCE ALIGNMENT

There are numerous applications for multiple sequence alignment, ranging from finding homologous regions in three or more sequences to finding new protein sequences in a protein family. One can easily perform multiple sequence alignment using several pairwise sequence alignments, but this solution is a time- and space-consuming one. Today, some novel methods can be used to align multiple sequences directly using well-known methods such as artificial neural networks and hidden Markov model approaches.

### 5.3.1 DYNAMIC PROGRAMMING

We can easily use the pairwise alignment dynamic programming method to align three or more sequences. We can increase the dimensions of our scoring matrix for multiple alignments and use similar initialization and filling methods for three or more dimensional matrices. If the number of sequences arises to 10 or more, dynamic programming cannot be suitable even for laboratory uses.

Because of the complexity of dynamic programming, some heuristic methods are used by researchers who want to align multiple sequences. Two families of methods can be distinguished: progressive and iterative methods.

### 5.3.2 PROGRESSIVE METHODS

In progressive methods, the pairwise alignment is used in some way to achieve the multiple aligned sequences. In this approach, two random sequences are chosen and aligned by a pairwise alignment method (e.g., dynamic programming). The resulting alignment is then compared with the third sequence, and the consensus alignment resulting from the three becomes the representative of these three sequences. The algorithm runs until all of the sequences are aligned successfully into this consensus sequence. As it can be seen, the choice of first two sequences is important in this method. So, one can use all of the pair sequences as an initiator and align them by progressive methods. The best scoring alignment of all sequences is chosen, then this alignment as a sequence for second (and subsequent) alignments. The program runs until all of the sequences have been used.

The progressive method can be used to align local sequences as well as global sequences because it can align pairwise sequences locally and expand the alignment to all of the sequences.

Errors resulting from primary alignments propagate through subsequent alignments in progressive methods. This is the most important weakness of these methods. Using progressive methods, the final result of alignment is highly proportional to the number of alignments.

### 5.3.3 ITERATIVE STRATEGIES

Iterative methods realign subgroups of sequences repeatedly by using, for example, a genetic algorithm.

Approaches based on genetic algorithms start with a random definition of gap insertions and deletions and often use the alignment score as a fitness function that drives the artificial selection of relatively fit alignment candidates to be included in the next generation of the population. The definition pattern of gaps and the position of each sequence are allowed to mutate and align with other patterns. In this way, *an* optimal—but not *the* optimal—multiple alignment between sequences is obtained over time.

## 5.4 SUMMARY

The various strategies presented within this chapter all hinge on a single assumption: that the similarity in primary structure—be it in proteins or DNA—between two sequences is directly related to the similarity in both form and function. Whether the methods are exhaustive (dynamic programming) or heuristic (FASTA, BLAST), they all attempt to quantify the similarity between two arbitrary sequences in the hopes that this will allow us to infer properties of new strings from large databases of well-understood strings. Because the volume of data is significant, and grows very quickly, even small refinements to these alignment methods can yield significant improvements, ensuring that this will remain an active research area for biological and computational scientists for the foreseeable future.

## 5.5 PROBLEMS

1. Define dynamic programming method and describe how it is used for sequence alignment.
2. What are BLAST and FASTA? What are the main differences?
3. Compare and contrast pairwise and multiple sequence alignment.
4. What does a long diagonal of filled cells in a dot plot represent? How does the interpretation vary between a long diagonal on the identity line and a long diagonal located off the identity line?

# 6 Genomics and Proteomics

## 6.1 INTRODUCTION AND OVERVIEW

Genomics and proteomics—together simply referred to as the "omics" within bioinformatics—are the high-level, sense-making loops in which researchers seek to understand the interplay among genes and proteins, respectively. Omics are the studies that occur after sequencing is complete and may include inferences about expression, homology/conservation, and biological function. Consider a classroom full of students: Demographics—akin to sequencing—provides the summary characteristics of the individuals present, whereas sociology—here likened to omics—might study how these individuals self-organize to form cliques or other functional subgroups. In many ways, genomics and proteomics directly address the "so what?" question that arises from the time, effort, and expense of collecting large volumes of genetic and protein data.

The remainder of this chapter addresses genomics and proteomics in turn, presenting each with respect to its sequence data, its annotations, and discussion of various types of postsequencing (omic) analysis.

## 6.2 GENOMICS

Genomics is the study and analysis of a genome, the fully enumerated sequence of an organism's genetic material. This section describes genomics, beginning with a review of the genetic sequence data available and its annotations and proceeding to a review of the genomic analysis that can be applied to those data.

### 6.2.1 GENETIC SEQUENCE DATA

To sequence an organism's genome is to create, typically from clonal cells, one exemplar model of the entire ordered list of nucleotides in that organism. The human genome—for which the Human Genome Project announced their first working draft in 2000—contains approximately 3.1 billion base pairs. The Genomes OnLine Database lists 843 completed genomes in their database* (mainly bacteria), more than 3,000 ongoing genomes, and 130 metagenomes (described in a subsequent section).

The methods for sequencing DNA appear in earlier chapters, allowing us to focus here on ways in which these collected data can be useful.

There are many repositories for genome (sequence) data, including

Genomes OnLine Database: http://www.genomesonline.org/
The Sanger Institute: http://www.sanger.ac.uk/

---

* http://www.genomesonline.org/ as of the August 20, 2008, update.

The J. Craig Venter Institute: http://www.tigr.org/
The Human Genome Sequencing Center at the Baylor College of Medicine: http://www.hgsc.bcm.tmc.edu/seq_data/

### 6.2.2 GENETIC ANNOTATIONS

Once sequencing is complete, the initial round of research and analysis consists of trying to identify functional elements within the sequence. These functions may be divined experimentally or—as the number of databases increases—by comparing the candidate sequence with similar sequences that have already been explored, hoping that a strong similarity between sequences suggests similar function.

These inferences, independent on how they are generated, are often used to annotate the raw sequence data by point or region (allowing them to be searched and reused) and may include

The demarcation of genes within the sequence
The identification of coding and noncoding regions of DNA
The enumeration of proteins that are expressed from the given subsequence
The association between the gene(s) within a subsequence and phenotypes they influence

There are extensive repositories for annotations of primary sequence data, including

ENSEMBL, "a joint project between EMBL-EBI (The European Molecular Biology Laboratory's European Bioinformatics Institute) and the Sanger Institute": http://www.ensembl.org/index.html; there is an entire family of subprojects that are based on ENSEMBL infrastructure.
The Gene Ontology: http://www.geneontology.org/GO.current.annotations.shtml
SEED: http://www.theseed.org/
NIH's National Human Genome Research Institute's Encyclopedia of DNA Elements (ENCODE): http://www.genome.gov/10005107

### 6.2.3 GENOMIC ANALYSIS

Understanding gene networks is one of the principal applications of genomic analysis. Although there is still interest in simple Mendelian mechanisms—one gene controls one phenotypic property—most genes contribute to phenotype much less directly and are only effective in conjunction with the expression patterns of other genes. Modeling gene networks, and inferring expression pathways, are subjects of another chapter within this text.

It is important to note that these networks may not be constructed of simple, direct links between a pair of cells of the form "gene A becomes active at time T, which causes gene B to become active at time $T + 1$." In fact, the expression pattern of one or more genes can influence the expression of a large number of genes; this network of influences is termed *epistasis*. Contrast with this the term *epigenetics*, which refers to the differing expression levels in genes across generations that result in new phenotypes without changing the underlying genetic structure of the organisms. Dmitri Belyaev's experiments with foxes in Siberia showed how artificially selecting animals that were more docile produced offspring in subsequent generations with phenotypes that included properties associated with domesticated dogs: floppy ears, spots, and curly tails. Adrenaline, it is hypothesized, participates in some indirect pathway among genes, so that decreasing the level of this hormone over generations altered the expression patterns of genes without altering the genes themselves. That is remarkable because it suggests that the phenotypes commonly observed are the result of merely one expression pattern within an organism's DNA.

The advantages of genomic analysis—a holistic study of the genes—are clear: Better understanding of genes and their interactions should allow for better understanding (and control) of regular cell processes and aberrant cell processes (diseases). This opens doors to improved (perhaps even personalized) medical treatments and pharmaceuticals.

### 6.2.4 METAGENOMICS

In much the same way that genomics recognizes the interdependence of genes to support an organism, metagenomics recognizes that an organism's success is often linked to other organisms. Metagenomics is the study of the metagenome—the community of individual genomes—present in an environmental sample, such as seawater or soil. The number of different member DNA strands in such a sample is often extremely large, so large-scale sequencing produces a huge amount of data. The resulting challenge is to analyze not only what the interplay may be among genes within a single genome, but to understand how genes influence each other among different organisms within the same local environment. This is not so much population genetics as it is a study into the interaction between organisms that both compete and cooperate: Within a single environment, for example, there may be both competitors for scarce resources as well as symbiotes that support each other by participating in different phases of a longer chain of chemical processes.

## 6.3 PROTEOMICS

Proteomics is the study and analysis of the proteome, the fully enumerated collection of an organism's expressed proteins. Consider that, although DNA contains the base information for the pattern of the cell's life, it is only through the expression of these genes into proteins that any of the cellular processes are accomplished, and yet, this translation of base information into biological function is far from direct: The genomics described in the preceding section—complicated enough—becomes increasingly complicated when it is described at the level of expressed proteins. These are some of the core issues that proteomics addresses.

This section describes proteomics, beginning with a review of the protein data available and its annotations, and then proceeds to a review of the proteomic analysis that can be applied to those data.

### 6.3.1 Protein Sequence Data

Alternate gene splicing, discussed elsewhere within this text, is the mechanism by which a single eukaryotic gene can produce multiple proteins, so many proteins, in fact, that even when it is possible to identify all of the introns and exons explicitly, it can be exceedingly difficult to quantify the relative amount of each of the possible proteins actually expressed. This means that for practical purposes, a proteome is more likely to be identified using wet laboratory methods than by using computational methods. This also means that, because the number of proteins dwarfs the number of genes, that proteomic analysis is correspondingly more complicated.

Some of the databases that store primary sequence data for proteins include

The Worldwide Protein Data Bank: http://www.wwpdb.org/
Uniprot: http://www.uniprot.org/

Resources that help expose assembled proteomes include

The Plant Proteome Database: http://ppdb.tc.cornell.edu/
NCBI UniGene: http://www.ncbi.nlm.nih.gov/sites/entrez?db=unigene
The Global Proteome Machine Proteomics Database: http://gpmdb.rockefeller.
    edu/
The Resource Center for Biodefense Proteomics Research: http://www.
    proteomicsresource.org/Resources/Catalog.aspx

### 6.3.2 Protein Annotations

Recall that there are four types of structure for a protein:

Primary: the sequence of amino acids that comprise the protein
Secondary: the internal, 3-D structures—such as alpha helices and beta sheets—that contribute to the overall 3-D folded conformation of the protein
Tertiary: a final, folded conformation of the protein
Quaternary: the contribution of the protein to a larger complex of proteins

All of these structure levels are subject to annotation, although the fourth might most properly be considered within the realm of proteomics (as it relates to the interactions among proteins). As researchers identify proteins in samples, these proteins are purified, sequenced, and analyzed to determine their folded conformation. These conformational data are also used to examine the homology and the stand-alone function of the protein within its biological context. The manner in which the proteins go about assuming their final folded conformation may also be subject to annotation.

These annotations are stored within one of many different databases, including

UnitprotKB: consists of two relevant subsystems: Swiss-Prot (manual annotations) and Translated EMBL (automated annotations); http://www.uniprot.org/
The Swiss Institute for Bioinformatics's Expert Protein Analysis System (ExPASy): http://ca.expasy.org/

National Center for Biotechnology Information's Entrez: aggregates protein
structure data from multiple other sites; http://www.ncbi.nlm.nih.gov/
entrez/query/static/advancedentrez.html

Georgetown University Medical Center's Protein Information Resource
(PIR): http://pir.georgetown.edu/

Protein Structure Initiative Structural Genomics Knowledgebase (PSI
SGKB): http://kb.psi-structuralgenomics.org/KB/

The EMBL-EBI: http://www.ebi.ac.uk/Databases/protein.html

The Gene Ontology: http://www.geneontology.org/GO.current.annotations.shtml

The Center for Biological Sequence Analysis at the Technical University of
Denmark: http://www.cbs.dtu.dk/databases/

### 6.3.3 PROTEOMIC ANALYSIS

Proteomics is primarily the study of the mechanisms by which proteins interact
within larger biological networks. This focuses the research task on understanding the interactions among pairs (or groups) of proteins. This is more complicated
than genomic analysis, in that the number of combinations is greater, yet simpler
inasmuch as a single protein-protein interaction is better defined than a single gene-
gene interaction.

One of the difficulties in analyzing protein interactions, of course, is correctly
identifying the context within which the proteins might interact: Given the massive
number of proteins that may be present within a single cell, under what conditions
will any two specific species be present to interact? Often, this question is addressed
top-down, that is, given an interesting function, such as the formation of actin fibers,
identify the concomitant proteins and determine which are likely candidates for
interaction with the previously identified key species.

Some of the protein-protein interaction databases include

The Biological General Repository for Interaction Datasets: http://www.the-
biogrid.org

The EBI's IntAct database: http://www.ebi.ac.uk/intact/site/index.jsf

The Database of Interacting Proteins: http://dip.doe-mbi.ucla.edu/

The Molecular Interaction Database: http://mint.bio.uniroma2.it/mint/Wel
come.do

The Protein-Protein Interaction Database: http://www.anc.ed.ac.uk/mscs/PPID/

The Munich Information Center for Protein Sequences' Mpact: http://mips.gsf.
de/genre/proj/mpact

The Microbial Protein Interaction Database: http://www.jcvi.org/mpidb/about.php

## 6.4 SUMMARY

Genomics and proteomics are indicative of the systems biology approach, in that
they are attempts to explain biological function and processes from a holistic point
of view. Whether this means examining gene networks or the more detailed protein interactions, the omics rely on very large data stores and automated search and
analysis methods to inform the life sciences. Better understanding of these processes

at both the genetic and protein level will lead directly to new medical treatments and pharmaceuticals and hopefully allow us to improve the human condition beginning inside our own cells.

## 6.5  PROBLEMS

1. Using some of the online resources you are aware of, identify a gene network.
   a.  What is its associated biological function?
   b.  Where do the data come from, and how were they derived?
2. There are fewer genes than proteins in eukaryotes, but each gene can encode multiple proteins, so which ought to be more complicated, genomics or proteomics?
3. Identify a protein with a well-established quaternary structure. What is its role? What does it accomplish in conjunction with other proteins? How was it first identified?

# 7 Methods for Identification of Differentially Expressed Genes or Proteins

## 7.1 INTRODUCTION AND OVERVIEW

In this chapter, we review the methods that detect genes differentially expressed across two sets of experiments. These two sets of experimental conditions can be two different laboratory settings, healthy versus diseased cases, or before and after introducing a drug/treatment. This chapter begins with some basis questions regarding the applicability of simple methods such as $t$ test and continues with more advanced methods.

## 7.2 WHY $t$ TEST IS NOT ENOUGH?

Recently, high-throughput technologies, in particular DNA microarray, have provided the means for simultaneous screening and analysis of thousands of genes. Although an enormous volume of data is being produced by microarray technologies, the full potential of such technologies cannot be accessed without the ability to sift through the noisy signals to obtain useful information. The large data sets produced by microarray technology have resulted in the need for reliable, accurate, and robust methods for microarray data analysis. A major challenge is to detect genes with differential expression profile across two experimental conditions. In many studies, the two sample sets are drawn from two types of tissues, tumors, or cell lines or at two time points during the course of biological processes.

The computationally simple methods used for such analysis, including the methods of identifying genes with fold changes (such as the popular log-ratio graphs), are known to be unreliable because in such methods the statistical variability of the data is not properly addressed. Although various parametric methods and tests such as two-sample $t$ test have been applied for DNA microarray data analysis, strong parametric assumptions made in these methods as well as their strong dependencies present in large sample sets restrict the reliability of such techniques in typical DNA microarray studies. One such assumption made in $t$ test is that all variables are Gaussian distributed, which may not be true for many DNA microarray applications.

## 7.3   MIXTURE MODEL METHOD

The main goal in analyzing microarray data is to determine the genes that are differentially expressed across two types of tissue samples or samples obtained under two experimental conditions. Mixture model method (MMM) is a nonparametric statistical method often used for microarray processing applications. Although MMM is one of the most popular methods used for identification of differentially expressed genes, it is known to overfit the data if the number of replicates is small.

It is claimed and argued that the new extensions of the MMM are among the best methods producing biologically meaningful results. The major disadvantages of the above-mentioned methods, especially the MMM, include the lack of repeatability of the results under different runs of the algorithm and the sensitivity of the algorithm on parameter initialization. A reliable microarray analysis method must be reproducible and applicable to different data sets under different experimental conditions. More specifically, an accurate microarray processing method is expected to pinpoint, with a relatively high degree of accuracy and robustness, genes with elevated expression levels that are related to the experimental condition in all runs.

### 7.3.1   BASIC FORMULATIONS

We start this section with the description of MMM technique. Consider $Y_{ij}$ as the expression level of gene in array $j$ ($i = 1,\ldots, n; j = 1,\ldots, j_1, j_1 + 1,\ldots, j_1 + j_2$), where the first $j_1$ and last $j_2$ arrays are obtained under two conditions. A general statistical model for the resulting data is

$$Y_{ij} = a_i + b_i x_j + \varepsilon_{ij} \tag{7.1}$$

where $x_j = 1$ for $1 \leq j \leq j_1$ and $x_j = 0$ for $j_1 + 1 \leq j \leq j_1 + j_2$. In addition, $\varepsilon_{ij}$ is a random error with mean 0. From the above formulation, it can be seen that $a_i + b_i$ is the mean expression level of the first condition and $a_i$ is the mean expression level of gene $i$ in the second condition. The method requires that both $j_1$ and $j_2$, the number of data sets for each experiment condition, be even.

The $t$ test statistic type scores are calculated on the preprocessed data. Here, $a_i$ is a random permutation of a column vector that contains $j_1/2$ 1's and $j_1/2 - 1$'s and $b_i$ contains $j_2/2$ 1's and $j_2/2 - 1$'s.

$$z_i = \frac{(Y_{i(1)}a_i / j_1) + (Y_{i(2)}b_i / j_2)}{\sqrt{v_{i(1)} / j_1 + v_{i(2)} / j_2}} \tag{7.2a}$$

$$Z_i = \frac{\bar{Y}_{i(1)} - \bar{Y}_{i(2)}}{\sqrt{v_{i(1)} / j_1 + v_{i(2)} / j_2}} \tag{7.2b}$$

Since the data are not assumed to be normally distributed, the distribution functions $f_0$ and $f$ are estimated. The null distributions, $f_0$ and $f$, are estimated directly in a nonparametric model for gene expression data.

$$f_0(z;\Phi_{g_0})=\sum_{i=1}^{g_0}\pi_i\phi(z;\mu_i,V_i) \tag{7.3a}$$

$$f(Z;\Phi_g)=\left(1-\sum_{\gamma=1}^{g}\pi_\gamma\right)f_0(Z;\Phi_{g_0})+\sum_{i=1}^{g}\pi_i\phi(Z;\mu_i,V_i) \tag{7.3b}$$

where $\phi(y;\mu_i,V_i)$ symbolizes the normal density function with mean $\mu_i$, variance $V_i$, and the mixing proportions $\pi_i$ define the linear combination of the normal basis function. We use $\Phi_{g_0}$ to represent all unknown parameters $\{(\pi_i, \mu_i, V_i):i = 1,\ldots, g_0\}$ in a $g_0$-component mixture model. The number of normal basis functions, i.e., $g_0$, can be estimated by the expectation maximization (EM) algorithm, which maximizes the log-likelihood function to obtain the maximum likelihood estimation of $\hat{\Phi}_{g_0}$.

$$\log L(\Phi_{g_0})=\sum_{j=1}^{N}\log f_0(z_j;\Phi_{g_0}) \tag{7.4}$$

Within $K$ iterations, the EM algorithm is expected to find the local maxima for all unknown parameters. It is recommended to run the EM algorithm several times with various random starting parameters and choose the final estimate as the one resulting in the largest log likelihood. As mentioned above, using random starting points causes the result of the MMM to be unstable and avoids reproducibility of the results. More specifically, in each run, the MMM algorithm may give a different number of expressed genes, which is not desirable in biological studies. This issue will be addressed later in this chapter.

After finding the optimized $\hat{\Phi}_{g_0}$ for different $g_0$'s, the algorithm selects the suboptimal $g_0$ corresponding to the first local minimum of BIC or AIC.

$$\text{AIC}=-2\log L(\hat{\Phi}_{g_0})=2v_{g_0}, \tag{7.5a}$$

$$\text{BIC}=-2\log L(\hat{\Phi}_{g_0})=v_{g_0}\log(N) \tag{7.5b}$$

where $v_{g_0}$ is the number of independent parameters in $\Phi_{g_0}$. Then, the algorithm uses the resulting $g_0$ as the number of normal functions to fit $f_0$. The same procedure is performed to estimate the suboptimal number of normal functions to estimate $f$. As mentioned above, with the fixed number of normal functions, the parameters of functions $f$ and $f_0$ are iteratively updated for a number of iterations. When the iterations are terminated, the likelihood ratio parameters are estimated based on the final estimations of $f_0$ and $f$:

$$\mathrm{LR}(Z) = f_0(Z) / f(Z) \tag{7.6}$$

For a threshold value $s$, if $\mathrm{LR}(Z) < s$, then the gene is identified to have significantly altered its expression in two experiments. It is possible to determine the rejection region numerically, i.e., for any false-positive rate $\alpha$, the threshold value $s = s(\alpha)$ can be calculated from the following integral:

$$\alpha = \int_{\mathrm{LR}(z)<s} f(z)\,\mathrm{d}z \tag{7.7}$$

In the literature of microarray processing, $\alpha = 0.01$ is often used as the genome-wide significant level, so the gene-specific significance level is $\alpha^* = \alpha/(2n)$.

For the cases where $j_1 \geq j_2$ but $j_1 < 2j_2$, $j_1$, observations drawn under condition one are split into two equally sized parts to calculate $\overline{Y}_{i(1a)}, v_{i(1a)}, \overline{Y}_{i(1b)}$, and $v_{i(1b)}$, respectively. To calculate and $\overline{Y}_{i(2)} v_{i(2)}$, about $j_1/2$ observations are drawn under condition 2. Although this modification can address the differences in the distributions of $f$ and $f_0$, the stability of the parameter estimation step still remains a major problem. The main difference between the conventional MMM and its recent extensions is that the conventional MMM disregards the fact that the true distribution of $z$ (the statistical variable under study) may be different from the null distribution of the statistics $Z$ (as defined below). This assumption can potentially lead to invalid inference.

### 7.3.2 More Advanced Versions

A concern over all existing MMM-based methods that greatly affects the results is associated with the way mixed distributions are estimated. In the MMM, expectation maximization (EM) algorithm is often used to optimize the parameters of fitted mixture distribution functions of two $t$ statistic–type scores related with gene expression level. Starting the EM algorithm with random values as the parameters of the normal basis functions to estimate distributions makes the results depend highly on the exact initialization and always makes variations in the results. On the other hand, if all parameters of the normal functions in the mixture model distributions are set without iterative optimization, the set values may never result to an accurate model of the data set in hand. We also describe a modified version of MMM (K5M hereafter) that combines $K$-mean clustering and the EM estimation to not only optimize most of the parameters with the EM iteratively but also apply $K$ means to optimize other sensitive parameters to ensure complete reproducibility of the algorithm. The experimental results indicate superior robustness of K5M algorithm compared with the conventional MMM and other recently introduced extensions of the MMM.

To address the stability and reproducibility of the MMM, this modified approach to MMM estimates the distribution function of $z$ by using mixture of normal distributions in a stable and reliable way. The following observations made in the experimental study of the MMM for gene expression analysis were the main motivations for the changes to the MMM:

1. The observed variations in the parameter estimation process in some versions of the MMM can be attributed to the algorithm's attempt to iteratively update the means and variances of the normal distributions using often noisy data. In experimental studies, often the direct observation of the data reveals specific points where centers (means) can be positioned and the scattering patterns that can give reliable estimates on the variance of each cluster. However, the iterative updating of model parameters with noisy data and based on some random starting points often misses the true optimal points and even creates variations and fluctuations in parameter estimation in many runs.

2. Even when variations do not occur, two runs of the algorithm can result in significantly different estimations of $f$ and $f_0$. This, in turn, results in lists of differentially expressed genes in different runs. More specifically, a set of two typical runs of the algorithm on the same data set can result in two lists that are very different both in the number of the genes as well as the exact genes picked up by the algorithm. In a typical study of the conventional MMM, two runs with the same algorithm (on the same data) resulted in lists whose size varies between 50 and 200.

3. The literature of other areas of research utilizing normal basis function for estimation including neural networks indicates that to have more robustness in different runs and have reproducible results, the means and variances of the basis functions must be estimated and fixed during the iteration on the coefficients. This is because updating means and variances makes the estimation process a nonlinear one that is highly sensitive and very likely to become unstable. However, when updating the values of coefficients only, the problem is reduced to a reliable linear estimation that is much more robust and stable.

4. Based on the observations mentioned above, in our proposed method, finding the distribution of $z$ is regarded partially as a clustering problem, i.e., the means and variances of the normal distributions are estimated as the prototypes of a clustering step. Specifically, if $z$ is distributed in a one-dimensional space, wherever there is a mass of $z$, there is a cluster with mean $\mu_i$ and variance $V_i$, which are identified by the members of that cluster. Hence, applying a clustering method is capable of estimating the means and variances of each normal distribution. The key is to use a simple clustering technique such as $K$ mean to estimate the mixture distributions $f_0$ and $f$ based on $K$ normal distributions. Although the algorithm can use $K$ means to find the optimal values of means and variances, the coefficients $\pi_i$'s need to be optimized using an optimization process such as the EM.

Based on the above discussion, the K5M algorithm can be described in the following three steps:

Step 1:   Using BIC, find the suboptimal number of normal distributions for both $f$ and $f_0$ (as described above). These optimal numbers are then used as the number of clusters in $K$-means technique.

Step 2:   Using $K$-means clustering technique, for both $f$ and $f_0$, find the best mean $\mu_i$, and variance $V_i$ for all clusters.

Step 3:   With the obtained values of $\mu_i$, $V_i$, and using the EM algorithm, iteratively update the values of the optimized $\pi_i$ for all clusters (both $f$ and $f_0$), i.e.,

$$\pi_i^{(k+1)} = \sum_{j=1}^{N} \tau_{ij}^{(k)} / N,  \tag{7.8a}$$

$$\tau_{ij}^{(k)} = \frac{\pi_i^{(k)} \phi(z_j; \mu_i^{(k)}, V_i^{(k)})}{f_0\left(z_j; \Phi_{g0}^{(k)}\right)}  \tag{7.8b}$$

## 7.4   GENES INVOLVED IN LEUKEMIA: A CASE STUDY

In this section, first, an application and its corresponding data set are described, and then the results produced by K5M is compared with the other MMM-based methods on the data set.

In this section, we apply the nonparametric MMM method with and without the proposed modifications to the leukemia data presented by Golub et al. (1999). The objective of this application is to identify the most important genes involved in the development of the different types of leukemia. The data set used for this analysis includes 27 acute lymphoblastic leukemia (ALL) samples and 11 acute myeloid leukemia (AML) samples for 7,129 genes. The main goal is to find genes with differential expression between ALL and AML cases. A second goal is to compare the result of MMM (as introduced in Section 4) with K5M and test the robustness of K5M. The genome-wide significance level is chosen $\alpha = 0.01$. Each sample in the data set is preprocessed by subtracting its median and dividing the resulting variable by its quartile range (i.e., the difference between the first and the third quartiles) as suggested by Pan (2002).

Thomas et al. (2001) used known biological information to identify the most important genes in leukemia and provided biological justifications for these identified genes. They introduced 50 genes out of the identified genes as the most expressed and related genes to the disease, including 25 most expressed genes for AML and 25 for ALL. We treat the list of Thomas et al. as the biology knowledge base and compare the capabilities of the computational techniques to correctly identify the genes discussed in their study by processing the data set.

The MMM has identified 187 differentially expressed genes (Pan, 2002), among which the total of 39 genes are in the list of genes obtained by Thomas et al. (2001). The K5M algorithm determines 45 genes that are identified in the Thomas et al. list, i.e., the proposed algorithm successfully identifies 90% of the biological result. This means that K5M improved the detection of expressed genes 12% compared with the MMM, i.e., K5M identified more genes from the list of the 50 truly expressed genes identified by Thomas et al. (2001).

As the BIC suggested the optimum number of clusters $K = 4$ for the MMM, the K5M is applied with $K = 4$ also. Running K5M with different numbers of clusters leads to the different but reasonably similar results. As the number of the clusters increases, the number of expressed genes decreases. The K5M with $K = 3$ identifies the total of 185 differentially expressed genes, whereas with $K = 4$, a total of 58 genes are identified, however; the 58 genes found with $K = 3$ are the most expressed genes among 185 genes found by $K = 4$. This result shows the consistency of the K5M method.

To further compare the performance of the MMM and K5M on the leukemia data, the ROC curve is plotted based on false- and true-positive rates of the data set. The area under each curve is the measure of test accuracy. As can be seen in Figure 7.1, the area under the K5M curve is more than the area under the MMM curve; therefore, the K5M is providing a more accurate classification than the MMM.

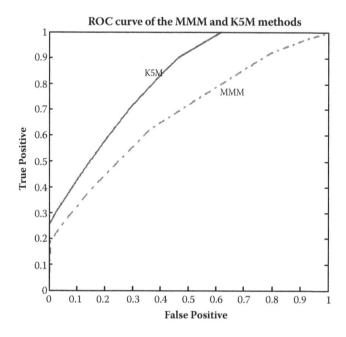

**FIGURE 7.1**    False- and true-positive rates.

## 7.5   SUMMARY

In this chapter, we described the disadvantages of $t$ tests and conventional MMM. We also described a method that improves the repeatability and robustness of the MMM by using the $K$-mean clustering method in estimating the distributions. This method finds the distribution of the variables partially based on a clustering procedure and an EM optimization process. The method is applied to analyze a microarray data set, the leukemia data set. The experimental results of K5M indicate 100% robustness and repeatability of the results in different runs and provide 12% improvement (compared with the MMM) in detecting the relevant genes in both studies.

## 7.6   PROBLEMS

1. Why is $t$ test not sufficient for DNA microarray processing? How does the MMM algorithm address this issue?
2. Describe the disadvantage of the MMM algorithm. Which disadvantages are addressed by K5M algorithm?

## REFERENCES

Golub, T., D. K. Slonim, P. Tamayo, C. Huard, M. Gaasenbeek, J. P. Mesirov, H. Coller, M. L. Loh, J. R. Downing, M. A. Caligiuri, et al., Molecular classification of cancer: class discovery and class predication by gene expression monitoring, *Science* 285 (1999): 531–537.
Pan, Wei, A comparative review of statistical methods for discovering differentially expressed genes in replicated microarray experiments, *Bioinformatics* 18 (2002): 4, 546–554.l.
Thomas, J., J. Olson, J. Tapscott, and L. Zhao, An efficient and robust statistical modelling approach to discover differentially expressed genes using genomics expression profile, *Genome Research*, 11 (2001): 1227–1236.

# 8 Binary and Bayesian Networks as Static Models of Regulatory Pathways

## 8.1 INTRODUCTION AND OVERVIEW

In this chapter, the methods to form binary networks as static regulatory pathways are reviewed. This chapter begins with some fundamental concepts and continues with the details of application in estimating gene/protein networks.

## 8.2 BINARY REGULATOR PATHWAYS

A challenging task in systems biology and bioinformatics is the estimation of regulatory interactions among the genes involved in a particular pathway using the data formed by high-throughout assays. In the recent years, several methods have been introduced to address this problem. More advanced family of methods dealing with this problem will be discussed in Chapter 9. In this chapter, we intend to focus on the simplest yet the most fundamental approach in addressing this issue, which is the use of binary networks.

The fundamental idea of binary networks is as follows. Consider genes A and B. If whenever gene A is active, it activates gene B, then the link (edge) between the two genes is set to 1. Similarly, if whenever gene A is active, it suppressed gene B, then the link between the two genes is set to −1. Finally, if the activation of gene A has no effect on gene B, then there is no effect from A to B. The link is often graphically represented by an arrow starting from A and ending at B. Since the value of the link can be either 1 or −1, such a network is sometime referred to as Boolean network.

## 8.3 BAYESIAN NETWORKS: ALGORITHM

Binary networks are popular tools in graph theory and decision-making systems in which a number of binary (0 or 1) nodes are connected via determinist links. Such networks are popular for systems biology and bioinformatics applications, such as discovering interactions among genes and gene products. However, binary networks are deterministic models, whereas DNA microarray data are noisy and clearly non-deterministic. This suggests that stochastic solutions, including Bayesian networks, might produce better models of such data. A Bayesian network, or belief network, is a probabilistic graphical model that represents a set of variables and their probabilistic independencies. Bayesian networks are able to handle incomplete data and discover causal relationships between variables and so are used to encode uncertain knowledge in many practical applications.

There are significant differences between Bayesian theory and the classical probabilistic approach. Suppose that we toss a coin: What is the probability of the outcome being a head? The Bayesian and classical approaches handle this question in different ways. The classical method simply calculates the probability that a coin will land heads. However, the Bayesian method generates an answer based on the confidence that the outcome will be a head, based on prior and observed facts.

A Bayesian network has two components: a directed acyclic graph (model selection) and a probability distribution (parameter estimation) representing a strong relationship among the variables. Nodes in the directed acyclic graph represent stochastic variables, and arcs represent directed dependencies among variables that are quantified by conditional probability distributions. Each directed edge connects two nodes; $E{\rightarrow}B$ represents an edge from parent $E$ to child $B$. Figure 8.1 shows three simple examples of qualitatively different probability relationships among three random variables.

The diagram in Figure 8.2 is an example of a joint distribution over a set of random variables—i.e., the probability of two events happening at the same time.

Thus, the network structure implies the conditional independencies:

$$I(E;B), I(R;A|E,B), I(C;R,E,B|A), I(R;A,C,B|E) \tag{8.1}$$

We denote $I(R;A|E,B)$ to mean $R$ is independent of $A$ given $E$ and $B$:

$$p(R;A|E,B) = p(R|E,B) \tag{8.2}$$

The network then implies the product form having the joint distribution:

$$p(C,A,R,E,B) = \prod_i p(node_i|parents_i)$$

$$= p(B)p(E)p(R|E)p(A|B,E)p(C|A) \tag{8.3}$$

That is, the joint probability of all of the variables is the product of the probabilities of each variable given the values of its parents.

## 8.4   APPLICATIONS AND PRACTICAL CONSIDERATIONS

The use of Bayesian networks is not limited to the discovery of the interaction among the genes using DNA microarray data. Recently, there have been many studies in which protein-protein interactions as well as protein-gene interactions have been explored using these computational tools.

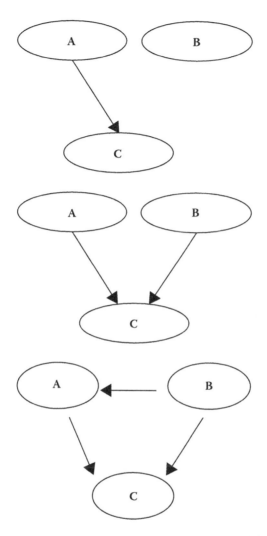

**FIGURE 8.1** Examples of different relationships among three variables. (a) $p(A,B,C)=$ $P(C|A) P(A)$, (b) $p(A,B,C)=p(C|A,B) p(A) p(B)$, and (c) $p(A,B,C)=p(C|A,B) p(A|B) p(B)$.

However, there are some practical considerations with regard to the use of Bayesian networks for bioinformatic applications that need to be discussed here:

- Estimation of probabilities from a few measurements available in DNA microarray data is often challenging and/or inaccurate. This is a systematic problem inherited in the nature of Bayesian methods and cannot be entirely addressed. However, some prior knowledge on the genes/proteins to be modeled can be used to improve the probability measures.
- When some of the probabilities discussed in Section 8.3 are close to 0.5, although the Bayesian model still reveals some information about the pathway, this information may not be biologically very useful.

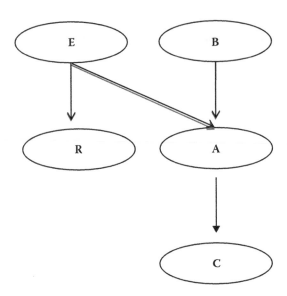

**FIGURE 8.2**   An example of a simple Bayesian network structure.

- The structure of regular Bayesian networks is designed to deal with static data, i.e., there is no mechanism to process time series and address the dynamic interactions occurring in time. This issue is addressed in a more advanced version of such models called dynamic Bayesian networks that are capable of dealing with time series.

## 8.5   SUMMARY

In recent years, there has been increased interest in the use of Bayesian networks in computational biology, and one popular application is the analysis of gene expression levels. Many diseases result from the interaction between genes, so understanding these interactions and the mechanism of gene expression is vital in developing new treatments. Bayesian networks are a promising tool for discovering gene structures and capturing gene interactions.

## 8.6   PROBLEMS

1. What are the main disadvantages of binary networks?
2. Can regular Bayesian networks be used to process DNA microarray time series? If not, what is the extension of these models that are capable of processing time series?

# 9 Metabolic Control Theory for Static Modeling of Metabolic Pathways

## 9.1 INTRODUCTION AND OVERVIEW

In this chapter, metabolic control theory and its application in analysis of static metabolic pathways are reviewed. This chapter begins with some fundamental concepts and continues with the details of application in estimating gene/protein networks.

## 9.2 BASIC IDEAS

As mentioned in previous chapters, a challenging task in systems biology is the inferring of regulatory interactions among the genes involved in a particular pathway using the data provided by high-throughput assays. Several methods have been introduced to address this challenging problem, among which some were mentioned in the previous chapters. Some of the methods designed to address this task include Boolean networks, Bayesian networks, linear models, dynamic Bayesian networks, techniques based on control theory, compartmental modeling using differential equations, full biochemical interaction models, and finally methods using metabolic regulation concepts. Although some of these techniques, such as Boolean networks, are too abstract and simplified, as discussed in previous chapters, others are computationally complex. The more complex models include techniques modeling all biochemical interactions among genes based on a large number of rather sophisticated differential equations.

The main advantage of the Boolean models is that they can handle a large number of genes and incorporate their binary interactions in the resulting network. However, the Boolean models express the activation of each gene as a binary value (1 or 0). This type of oversimplification results in a loss of important information about the pathway under study. Moreover, the Boolean models rely heavily on suitable choices of threshold values applied to convert real-valued microarray data to binary values (i.e., active or inactive). This thresholding step poses another restriction on the use of the Boolean models, as it is often difficult to determine an optimal threshold value to identify whether a particular gene is active. Since different genes can be active with different expression levels, the optimal choices of these threshold values can differ dramatically from one gene to another.

Models attempting to incorporate detailed biochemical interactions among the genes in a pathway are often limited to discovery of gene networks with a small number of genes. This is because these models require the estimation of a large number of parameters from a small set of microarray data. In other words, the statistical signal processing theory states that a model with many estimated parameters may not be reliable. A model with many parameters trained by only a few data points can simply overfit the training data. Some detailed models based on Bayesian networks theory are also known to suffer from the complications involved in estimating a large number of parameters.

The majority of gene regulatory models can be considered as specialized versions of the reverse engineering approach. These methods attempt to avoid oversimplification of the problem while obtaining more realistic models. A group of such methods have been specialized to process DNA microarray data containing mRNA of the genes before and after the perturbation of the biological system under study. Concepts in both control theory and metabolic control theory are utilized to quantitatively model the effects of the changes in the expression value of one gene on the expression level of other genes in the same pathway. Concepts of metabolic control analysis (MCA) are used to process the variations in expression value of the genes before and after pathway perturbations to estimate the regulatory network. Metabolic control uses control coefficients to determine the relationship among elements with respect to biochemical reactions and the state variables of the system.

The network created by almost all methods described above will include some direct links between each pair of genes but fail to discover indirect interactions between the two genes (i.e., interactions between two genes via intermediate genes). To see this disadvantage more clearly, consider the simple network shown in Figure 9.1.

As it can be seen in Figure 9.1, the network explicitly describes the "direct" effect of gene C on gene B with the direct links between these two genes. However, the network also indicates that gene C has some "indirect" effects on gene B through genes D and A. Estimating the overall interactions among a pair of genes (both direct and indirect) plays an important role in many practical biology studies. For instance, one

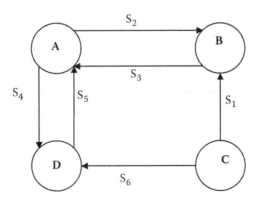

**FIGURE 9.1**  A typical gene regulatory network.

of the most important problems in designing gene knockout experiments is to establish a systematic procedure to identify the most informative gene(s) to be deleted in the following knockout experiment.

The main method described in this chapter, which is based on a set of techniques presented by Darvish and Najarian (2006), is a stepwise process that combines the concepts of metabolic control, graph, and control theories to form gene regulatory pathways.

Another advantage of the methods discussed in this chapter is incorporating gene-protein and protein-protein interactions to improve the models created based on DNA (or mRNA) microarray data. Specifically, a main disadvantage of some of the pathway identification methods is failing to properly incorporate all high-throughput data in the process of modeling; i.e., many DNA microarray processing methods disregard important gene-protein interactions while forming the pool of regulator genes.

## 9.3 MAIN CONCEPTS IN METABOLIC CONTROL THEORY

First, a method is introduced to determine the genes that have the most effect on the main regulatee gene(s) of the pathway under study and then identify the most correlated genes to the regulatee gene(s). In this process, the pool of selected regulator genes is not restricted to the genes whose mRNA expressions have highest correlations with the regulatee gene(s). Instead, the information on gene-protein interactions is considered in the process of choosing the relevant genes. Specifically, if a protein encoded by a gene is known to affect/alter the expression level of the regulatee gene(s), the gene encoding this protein is added to the pool of regulator candidates. In addition, a pruning technique is used to eliminate the effect of the measurement noise in forming weak links in the predicted pathway. In the next step, the links among the regulator genes are identified using the concepts in metabolic control theory. The resulting network is then analyzed using the Mason method to obtain the overall effects of each pair of genes on each other. In particular, the method generalizes the ideas of active learning of a network to compute the effect of each gene on each other. To address the direct and indirect interactions among the genes, a control theoretic algorithm based on Mason's rule is applied to estimate the interactions of each pair of nodes (genes) considering all pathways in the network connecting these two nodes. Using this quantitative approach, the method identifies the genes having the most effect on the regulatee gene(s) in the pathway. Then, if the genes with highest effects on the regulatee genes have not been perturbed yet, the method introduces these genes as the best candidate genes to be deleted in the next knockout perturbation experiment.

### 9.3.1 STEP 1

The first step in estimation of regulatory networks is identification of the genes of interest (i.e., regulatee genes), i.e., the goal is to find a pool of potential regulators for these target genes. To do this, a set of correlation tests based on the pathway perturbation experiments is performed to find the most correlated genes to the main gene of the pathway (e.g., Gal2 in the example provided later in this chapter). The

Pearson correlation measure, which has proved to be useful in finding genes with similar expression patterns, is often applied for this task; however, other measures such as concordance, mutual information, and Spearman correlation can also be used. The threshold on the correlation filter works on the absolute value of correlation to allow both correlated and anticorrelated genes to pass through the filter. Adding the anticorrelated genes might cause some confusion, as some genes from distinct antagonist pathways, which oppose the pathways of interest, may also pass through the absolute-value correlation filter. This is why the annotations of all genes with negative correlation are checked to analyze the involvement of the identified genes in the biological pathway under study. In this step, the genes that are known to have gene-protein interactions with the genes discovered through the correlation test are also explored. Specifically, other genes whose gene products (e.g., protein) are known to have some regularity effect on the genes of interest but are not passing through the correlation filter are added to the final list of the genes involved in the pathway. The main challenge for this process is the availability of such gene-protein interaction databases as discussed later.

The exact correlation measure used in this method is defined as follows:

$$\rho(mRNA_i, mRNA_j) = \frac{Cov(mRNA_i, mRNA_j)}{\sigma_{mRNA_i} \sigma_{mRNA_j}} \tag{9.1}$$

where $mRNA_i$ and $mRNA_j$ represent a vector containing the mRNA expression levels of genes $i$ and $j$ in different knockout experiments, respectively. In addition, $\sigma_{mRNA_i}$ and $\sigma_{mRNA_j}$ are the standard deviations of $mRNA_i$ and $mRNA_j$, and:

$$Cov(mRNA_i, mRNA_j) = E[mRNA_i mRNA_j] - E[mRNA_i]E[mRNA_j] \tag{9.2}$$

where $E(x)$ is expectation of $x$.

The method also searches the gene-protein interaction databases to find the genes whose protein products are known to regulate the regulatee gene(s). One major issue with inclusion of gene-protein interactions in this step is the lack of reliable and comprehensive gene-protein interaction databases. This shortage of data is primarily because of the very high cost of producing databases to express gene-protein interactions. In the galactose pathway identification example discussed in this chapter, a database is used that provides a relatively large list of protein-protein interactions to explore potential interactions on the gene-protein level. A major problem with inferring the protein interaction data to gene regulatory network is that there are many interactions between proteins that are not related to gene regulation, and as a result, a protein-protein interaction does not necessarily identify a gene regulation process. However, as it has been studied in many recent works, protein-protein interaction enhances the chance of regulation between a pair of genes. Therefore, in this chapter, since we do not have access to a reliable database of gene-protein interactions, we followed recent works, such as those of Ideker et al. (2001) and Ito et al. (2000), which use protein-protein interaction data to add another regulator gene to the pool of regulator genes of the main genes in the galactose pathway, as discussed later.

It has to be reemphasized that protein-protein databases such the database used in this study are not the best options for our algorithm; however, until more reliable gene-protein interaction databases are formed, incomplete protein-protein interaction databases could work as interim solutions.

### 9.3.2 STEP 2

Next, the interactions among the genes in the final list of regulators are examined to predict a quantitative model of the pathway. In this step, only the regulator genes identified by the previous step are included in the pathway formation process. To quantitatively explore the effects of the genes on each other, two important concepts used in MCA, i.e., co-control coefficient (Hofmeyr et al., 1993) and the regulatory strength (Jensen and Hammer, 1998), are estimated. A co-control coefficient expresses how the relative values of two variables change when a single parameter is perturbed, i.e., for genes $i$ and $j$:

$$O_{j,m}^{i,m} = \frac{\partial[mRNA_i]/[mRNA_i]}{\partial[mRNA_j]/[mRNA_j]} \tag{9.3}$$

where $m$ introduces that gene $m$ has been perturbed and $\partial$ is the differential operator. The regulatory strength, $R_{j,m}^{i,m}$, measures the degree at which the perturbation in a variable is propagated to another through a particular pathway, i.e.,

$$R_{j,m}^{i,m} = \varepsilon_{mRNA_i}^{v_j} C_{v_j}^{mRNA_j} \tag{9.4}$$

where $\varepsilon_{mRNA_i}^{v_j}$ is the elasticity of $v_j$ by $mRNA_i$, i.e., a local property of an isolated enzyme that expresses how its rate varies with the concentration of any metabolite that affects it, and $C_{v_j}^{mRNA_j}$ is the concentration-control coefficient of the transcription reaction $j$ on $mRNA_j$. These coefficients express the induction of changes in a variable (e.g., mRNA concentration) on another. Using Equation (9.4) to compute regulatory strengths may not be possible in many practical applications including the galactose pathway prediction discussed later in this chapter. This is because of the fact that microarray experiments do not allow the calculation of both the elasticity and control coefficients from data. Specifically, because of complications involved in calculating elasticity, it is more desirable to calculate the regulatory strengths directly from co-control coefficients to bypass the need to calculate elasticity (de la Fuente et al., 2002). This limitation, i.e., the complications involved in estimating elasticity from microarray data, is not specific to our approach and has been reported in almost all MCA formulations used for microarray data processing. The other issue is the continuous nature of the above equations, i.e., to implement the differentiation operator heavily used in the above formulation, one needs to have all variables (e.g., concentrations) as continuous functions. However, microarray data are merely sampled data and cannot provide a continuous range of readings for the variables. This limitation calls for the use of a discrete estimation of the above-mentioned variables and formulas.

The two issues discussed in Step 2 can be addressed as follows. First, in dealing with the discrete nature of the microarray data, one needs to estimate continuous differentiations in Equation (9.3) using discrete variables, i.e.,

$$O_{j,m}^{i,m} = \frac{\Delta mRNA_i / mRNA_i}{\Delta mRNA_j / mRNA_j} \tag{9.5}$$

The second issue is the calculation of the regulator strength directly from co-control coefficients. The following theorem relates regulatory strength to co-control coefficients.

## Theorem 1

A matrix of regulatory coefficients is equal to the inverse matrix of co-control coefficients.

The proof of Theorem 1 is given by (Hofmeyr and Cornish-Bowden, 1996) and is not repeated here. This theorem allows the calculation of regulatory strength using co-control coefficients. The matrix $O_i$ is the co-control coefficient matrix of gene $i$ that contains the control-coefficient between gene $i$ and each of the regulatee genes for different perturbations.

$$O_i = \begin{pmatrix} O_{i,1}^{1,1} & \cdots & O_{i,k}^{1,k} \\ \cdots & \cdots & \cdots \\ O_{i,1}^{n,1} & \cdots & O_{i,k}^{n,k} \end{pmatrix} \tag{9.6}$$

For instance, as can be seen in Equation (9.5), the element $O_{i,m}^{j,m}$ of matrix $O_i$ estimates the relationship between genes $i$ and $j$ when gene $m$ is perturbed. As such, this matrix gives the co-control coefficient between all genes and gene $i$. On the other hand, $R_i$ is the regulatory strength matrix of gene $i$. More specifically, the $i$th row of this matrix gives the regulatory strength of all genes on gene $i$, i.e., the element of the $i$th row and the $j$th column of $R_i$ gives the regulatory strength of gene $j$ on gene $i$.

Mathematically speaking, Theorem 1 states that the co-control coefficient matrix can be inverted to form the strength matrix for gene $i$, i.e.,

$$R_i = O_i^{-1} \tag{9.7}$$

The row $i$ of $R_i$ gives the regulatory strength of all regulator genes on gene $i$. Therefore, for each gene $i$, one can readily calculate the matrix $R_i$ and extract the $i$th row. This row corresponds to the regulatory strength of all genes of the network on gene $i$. Repeating the same process, i.e., computing $R_i$ for all genes ($i = 1,...,n$), the entire gene regulatory network can be obtained.

### 9.3.3   STEP 3

Because of the detailed structure of the Mason's method, the following section is dedicated to this step.

## 9.4 MASON METHOD

In this step, the overall interactions between each pair of genes through all the paths connecting them to each other are estimates. It has to be emphasized that each link value calculated in Step 2 gives the strength of the direct regulatory link between a pair of genes without considering the effect of indirect links between the two genes provided through other genes in the pathway. In Step 3, the overall interaction between each pair of genes is calculated in form of an overall gain using the Mason rule. These overall gains allow selecting the most promising regulators for the next knockout experiment. The most dominant regulators of the regulatee gene(s) are identified as the ones having the highest overall gains on the regulate gene(s) as discussed later. Mason's rule is a method in linear control theory as well as graph theory that can be used to obtain the overall gains between each pair of nodes in a control system. Mason's rule first finds all independent loops in each graph and then applies the identified loops to find all pathways (direct and indirect) between each pair of nodes in the network. Using all the identified pathways between each pair of nodes, Mason's rule then calculates the overall connection between each pair of nodes. Next, we briefly describe a version of the Mason's rule that is specialized for identification of gene regulatory pathways.

To best describe the method, a simple example is used. Consider the regulatory network shown in Figure 9.2. In this network, values $s_{11}, \ldots, s_{22}$ are the direct regulation effect (gain) from genes $a$ to genes $b$ and values $\Gamma_{22}$ and $\Gamma_{11}$ are the direct regulation effect (gain) from genes $b$ to genes $a$. For example, $\Gamma_{22}$ is the regulator effect of gene $b_2$ on gene $a_2$.

Assume that $b_1$ is the main gene of interest (i.e., regulatee) and we would like to find the overall effect of the gene $a_s$ on $b_1$. This means that we would like to estimate the overall gain $T = b_1/a_s$. As can be seen, $T$ is an overall gain that expresses how a change in expression value of gene $a_s$ can alter the expression level of gene $b_1$ considering all direct and indirect paths in the network connecting these two genes. This overall gain can be obtained using the Mason's rule as follows:

$$T = \frac{P_1[1 - \sum L_1^{(1)} + \sum L_2^{(1)} - \ldots] + P_2[1 - \sum L_1^{(2)} + \ldots] + \ldots}{1 - \sum L_1 + \sum L_2 - \sum L_3 + \ldots} \tag{9.8}$$

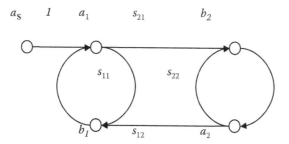

**FIGURE 9.2** A simple regulatory control network.

where
  $P_1, P_2,...$ are all pathways connecting the nodes
  $L_1$ is a first-order loop
  $L_2$ is a second-order loop
  $L_1^{(1)}$ is a first-order loop that has no common links with pathway $P_1$
  $L_2^{(1)}$ is a second-order loop that has no common links with pathway $P_1$

The first-order loop is defined as the product of gains of the branches encountered when starting from a node and moving along the loop in the direction of the arrows back to the original point. The second-order loop is the product of two nontouching first-order loops. For the network of Figure 9.2, following the procedure described above, the loops $L_1$, $L_2$ and the path $P_2$, identified with bold lines in Figure 9.3, are obtained.

The overall gain $T = b_1/a_s$ can then be calculated as follows:

$$T = \frac{b_1}{a_s} = \frac{s_{11}(1 - s_{22}\Gamma_{22}) + s_{21}\Gamma_{22}s_{12}(1)}{1 - (s_{11}\Gamma_{11} + s_{22}\Gamma_{22} + s_{21}\Gamma_{22}s_{12}\Gamma_{11}) + s_{11}\Gamma_{11}s_{22}\Gamma_{22}} \tag{9.9}$$

This procedural method allows obtaining the quantitative overall regulatory gain of each regulator gene on the regulatee(s) of the network and is instrumental in finding the most dominant regulators as the candidates for the future knockout experiments as discussed in the next step.

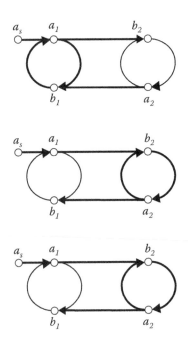

**FIGURE 9.3**    (a) Loop $L_1$, (b) loop $L_2$, and (c) path $P_2$.

Next, the developed gene-to-gene interactions are used to identify the most dominant regulators of the gene(s) of interest. The previous step provides the overall interactions between the gene(s) of interest and all other genes in the network. The most dominant regulator genes are the regulators having the largest overall regulation gain on the gene of interest. This is because a perturbation on these high-gain genes will result in the most significant perturbation of the regulatory pathway. Algorithm introduces these regulator genes as the best candidates for further knockout experiment, since deleting these genes would result to the most effective perturbation of the system. The importance of the active selection of the next genes to be knocked out becomes more evident when analyzing much larger regulatory networks containing up to a few hundred genes. In such large networks, to minimize the number of costly perturbation experiments, this method can provide a reliable model of the pathway without having to perturb all or most of the genes. The gene regulatory network is updated and improved throughout the iterative process of knockout experiment because after every new knockout experiment, the available model from the previous experiments is updated using the data generated in the new knockout experiment. In other words, the procedure starts with an incomplete data set that includes the perturbations of only a small number of genes, and then new optimal perturbations are planned to produce new data. The choice of the first few genes for perturbation can be made using prior biology information, correlation results, or both approaches. This process is continued to update the model with each new set of data generated at a consecutive knockout experiment.

A summary of the described method is given in Table 9.1.

The procedure described above starts from a small set of perturbation experiments and incrementally improves the model throughout an optimal design of the consecutive knockout experiments. Since the gene knockout and producing DNA microarray

---

**TABLE 9.1**

**The Summary of the Method**

Step 1:   Determine the most important regulatee(s).

Step 2:   Find the most correlated genes with the determined regulatee(s).

Step 3:   Apply the method based on MCA to obtain the initial network.

Step 4:   Use Mason's rule to find the most important regulator of the regulate.

Step 5:   Knock out the determined regulator in a new perturbation experiment and collect new DNA microarray data.

Step 6:   If there are more significant regulator genes to be knocked out, go to Step 4 and repeat the procedure; otherwise, go to Step 7.

Step 7:   Stop updating model.

---

are often time-consuming and expensive processes, conducting knockout experiments in large pathways with many regulator genes needs to be conducted using the experiment planning approaches such as the method. The procedural method described above suggests to biologists the best set of perturbation experiments to create a reliable model of the pathway as opposed to performing knockout experiments on all genes.

## 9.5  METABOLIC CONTROL MODEL FOR GALACTOSE REGULATION PATHWAY: A CASE STUDY

In this section, we apply the technique described above to predict the galactose regulation pathway and evaluate the performance of the technique in predicting the known galactose regulation pathway described by Ideker et al. (2001). The galactose regulation pathway is predicted based only on the mRNA expression level and protein-protein interactions, and then the resulting pathway is compared with the known facts about this pathway. The method is also applied to predict the potential interaction among genes that are not yet explored or verified by direct biological analyses.

Galactose pathway is a biochemical pathway in which galactose converts to glucose-6-phosphate. *Gal2* is the main gene in this pathway that transports galactose into the cell. *Gal4*, *Gal80*, and *Gal3* are the regulatory genes of this pathway that control the expression level of *Gal2*. The controller pathway regulates *Gal2* and, as a result, determines whether the pathway is on or off. As described by Ideker et al. (2001), in the absence of galactose, *Gal80* has a repression effect on *Gal4*. In addition, *Gal3* has an inhibitory (negative) regulatory effect on *Gal80* that causes *Gal80* to release its repression effect on *Gal4*. Our goal in this study is to form a model of the controller pathway by applying the method and test/verify the performance of the method in correctly predicting the biological facts known about the galactose pathway.

The only biological knowledge about galactose pathway used in the algorithm is the consideration of *Gal2* as the main gene of the pathway. First, the correlation test is performed to search for the genes that have high correlation with *Gal2* (as the main gene of interest). *Gal2* is chosen because of its central role in galactose regulation. To automatically find the main regulator genes in the pathway, first, we form a pool of candidate genes as potential direct or indirect regulators of *Gal2* by including all the genes in the known galactose pathway together with 490 other genes randomly selected from gene expression microarray data by Ideker et al. (2001). The data set given by Ideker et al. (2001) has several knockout experiments. In nine of these experiments, a gene from galactose pathway is deleted, and the expression values of all other genes are measured. This means that in evaluating correlation between any pair of genes, the experiments in which none of the two genes have been deleted can be used. After performing correlation test between *Gal2* and all 490 genes in the pool, the most correlated genes are selected as the regulators of *Gal2*. The genes selected by a correlation test are the ones for which the absolute value of correlation exceeds a threshold value of 0.85. Since literature on using correlation test for clustering of gene expression data reports experimentally optimal cutoff scores between 0.80 and 0.90, the threshold value of 0.85 was used in our analysis. These genes selected by this correlation filter are *Gal1*, *Gal10*, *Gal7*, *Gal3*, *Gal80*, and *Gal6*.

All these genes exist in the known galactose regulation pathway reported in the literature. This observation further indicates the suitability of the method used for selection of the regulator genes.

Up to this point, to find the most effective genes on *Gal2*, we considered only the correlation between all genes and *Gal2* in different knockout experiments. As mentioned before, mRNA expression levels may not sufficiently describe all the characteristics of the entire pathways, and to get a better model of the pathway, one also needs to consider other information such as gene-protein interaction (or even protein-protein interactions). Next, because of lack of a reliable gene-protein interaction database, we investigate the protein interactions of the genes in the preliminary list and see if other genes can be included in the list because of their high protein interactions with the genes in the preliminary list.

Examining the protein interactions of the genes passing through the correlation filter with other genes, it is observed that *Gal80* (one of the genes in our preliminary list) has protein interaction with *Gal4*, *Gal1*, and *Gal3*. Since *Gal1* and *Gal3* are already in the list, only *Gal4* needs to be added to the preliminary list to create the final list of genes in the galactose regulation pathway.

Next, we explore the interaction among the genes in our final pool of regulators. In the data set used for this study, all genes in the final set are perturbed. Therefore, we form the pathway using the following genes: *Gal1, Gal2, Gal3, Gal4, Gal6, Gal7, Gal10*, and *Gal80*. For each gene, we use eight experiments. In each of these eight experiments, one of the genes in the list is perturbed. Next, $R_{all}$ matrix whose $R_{ij}$ element indicates the regulatory effect of gene $i$ on gene $j$ is calculated. This matrix gives us the complete gene regulatory network.

$$
R_{all} =
\begin{bmatrix}
0.6 & -0.1 & 23.7 & 0.1 & -0.3 & 10.5 & -0.2 & -35.8 \\
-1.1 & 0.1 & -27.9 & -0.2 & 0.4 & -12.5 & 0.4 & 42.4 \\
-24.3 & 4.6 & -1010 & -6.3 & 8.9 & -42.9 & 11.5 & 1506 \\
7.3 & -1.4 & 307 & 1.8 & -1.7 & 129 & -3.6 & -458 \\
2.3 & -0.4 & 86.4 & -0.5 & -1.1 & 38 & -1.1 & -129 \\
-38.5 & 7.5 & -1628 & -9.8 & 18.2 & -690 & 18.7 & 2420 \\
37.9 & -7.2 & 1571 & 9.6 & -20.6 & 672 & -17.9 & -2338 \\
-0.2 & 0 & -8.5 & -0.1 & 0.1 & -3.7 & 0.1 & 12.7
\end{bmatrix}
$$

Next, the pruning step is performed. As can be seen, some elements of $R_{all}$ are very small in magnitude (in comparison with other elements) and do not introduce a meaningful link between two genes. To prune the matrix, we consider both row and column of the entries to be pruned or kept. In each row, elements whose absolute values are higher in comparison with other values in that row are preserved. In practical control applications, links whose absolute values are less than 33% of the entry with maximum absolute value of the row or column are replaced by zero. We loosely follow this experimental rule in pruning the above matrix. For example, in row 2, only the following elements are kept: 27.9, 12.5, and 42.4. After conducting the same pruning process for each row and column, the following pruned matrix is obtained.

$$
R_{prunedl} =
\begin{bmatrix}
0 & 0 & 23.7 & 0 & 0 & 10.5 & 0 & -35.8 \\
0 & 0 & -27.9 & 0 & 0 & -12.5 & 0 & 42.4 \\
-24.3 & 4.6 & -1010 & -6.3 & 8.9 & -42.9 & 11.5 & 1506 \\
0 & 0 & 307 & 0 & 0 & 129 & 0 & -458 \\
0 & 0 & 86.4 & 0 & 0 & 38 & 0 & -129 \\
-38.5 & 7.5 & -1628 & -9.8 & 18.2 & -690 & 18.7 & 2420 \\
37.9 & -7.2 & 1571 & 9.6 & -20.6 & 672 & -17.9 & -2338 \\
0 & 0 & -8.5 & 0 & 0 & 0 & 0 & 12.7
\end{bmatrix}
$$

Now, based on the elements of $R_{pruned}$, the network of the genes can be formed as shown in Figure 9.4.

Since our main focus is on the controller part of the galactose pathway, in the network of Figure 9.4, only genes included in the controller part of the galactose pathway have been shown. A comparison of the network in Figure 9.4 and the knowledge of galactose pathway reveal numerous similarities. As mentioned in the beginning of this section, from biological knowledge, it is estimated that *Gal3*, *Gal4*, and *Gal80* form the main regulatory pathway of *Gal2*, which in turns acts as the main switch in the pathway. In addition, biology studies report that *Gal80* has an inhibitory effect on *Gal4*, as reported in the literature. This biology observation clearly agrees with the negative link from *Gal80* to *Gal4* in Figure 9.4. All other links in Figure 9.4

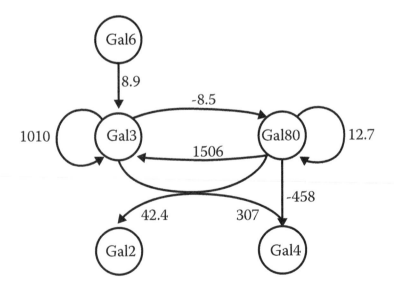

**FIGURE 9.4**  Resulting galactose pathway.

agree with the facts reported in literature. The self-regulation of genes in Figure 9.4 (e.g., self-link of *Gal3*) describes regulatory effects that are described in the literature as long loops containing other genes.

The observations on the discovered networks can be summarized as follows:

(a) As can be seen, the subnetwork of the genes forming controller part of the galactose pathway includes *Gal3*, *Gal4*, and *Gal80*. This matches the controller network as described in the literature.

(b) The obtained pathway indicates that *Gal80* represses *Gal4*. This fact matches with the observed biological knowledge.

(c) From Figure 9.4, it can be seen that when *Gal6* is active, it activates *Gal3*. This forms an active loop containing *Gal3* and *Gal80* that represses *Gal80*. This observation on the obtained model also matches with biology studies. In such a state, *Gal4* can also be active. Since *Gal6*'s expression level depends directly on the concentration of galactose, this loop forms a major role in the galactose regulatory pathway.

(d) Ideker et al. (2001) speculated that *Gal6* might have an effect on this regulatory network. This hypothesis is also quantitatively supported by the predicted link between *Gal6* and *Gal3* in our predicted pathway.

Next, we apply the Mason's rule on the detected galactose pathway and find the most dominant regulators of the main regulate of the pathway. But first, to show an example of the application of Mason' rule on galactose pathway, we obtain the overall effect of *Gal6* on *Gal4*. Figure 9.5 shows path $P_1$ in the pathway of Figure 9.4. The path gain can be calculated as

$$P_1 = (8.9)*(1010)*(-8.5)*(12.7)*(-458) = 444426000$$

Figure 9.6 shows path $P_2$ in the pathway of Figure 9.5. The path gain is then calculated as

$$P_2 = (8.9)*(1010)*(-8.5)*(-458) = 34994200$$

Path $P_3$ has been shown in Figure 9.7.

$$P_3 = (8.9)*(-8.5)*(-458) = 34647.7$$

Path $P_4$ has been shown in Figure 9.8.

$$P_4 = (8.9)*(-8.5)*(12.7)*(-458) = 440026$$

Now, the loop gains can be calculated as:

$$\Sigma L_1 = (-8.5 \times 1506) + 1010 + 12.7 = -11778.3$$
$$\Sigma L_2 = 12.7 * 1010$$

$$\sum L_1^{(1)} = 0$$

$$\sum L_2^{(1)} = 0$$

$$\sum L_1^{(2)} = 12.7$$

$$\sum L_2^{(2)} = 0$$

$$\sum L_1^{(3)} = 1010 + 12.7 = 1022.7$$

$$\sum L_2^{(3)} = 12.7 * 1010 = 12827$$

$$\sum L_1^{(4)} = 1010$$

$$\sum L_2^{(4)} = 0$$

Therefore, all transfer functions can be calculated. For instance, to calculate $T = (Gal4 / Gal6)$, we have

$$T = \frac{Gal4}{Gal6} = \frac{P_1 + P_2(1 - \sum L_1^{(2)}) + P_3(1 - \sum L_1^{(3)} + \sum L_2^{(3)}) + P_4(1 - \sum L_1^{(4)})}{1 - \sum L_1 + \sum L_2} = \frac{34119}{24605} = 1.39$$

This calculation results to an important observation that verifies our previous discussion. Here, *Gal6* was chosen as an example because, as can be seen in Figure 9.4, there is no direct link between *Gal4* and *Gal6*; however, using Mason's rule, the model predicts that *Gal6* has a measurable indirect effect on *Gal4*. To obtain the indirect regulatory effect between each pair of genes, this process must be repeated for all of the genes in the network. In particular, the overall gain between each regulator gene and the regulatee gene(s) must be calculated to identify the most dominant regulators of the network.

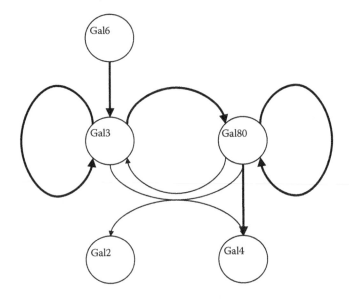

**FIGURE 9.5**  Path $P_1$ of the obtained pathway.

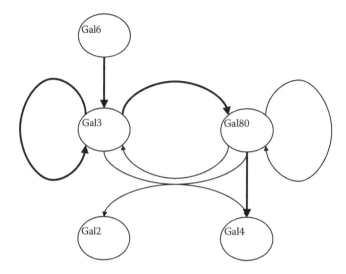

**FIGURE 9.6**    Path $P_2$ of obtained pathway.

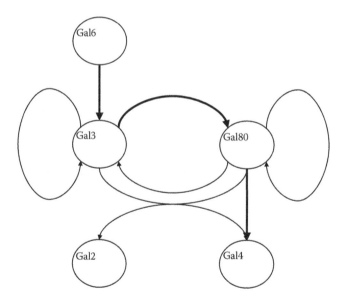

**FIGURE 9.7**    Path $P_3$ of obtained pathway.

In the next step, one must target the most important regulatee genes in the pathway from the prior biological knowledge and attempt to find the most important regulators of the regulatee gene(s). In this application, from biological knowledge, it is known that *Gal2* plays the role of the main regulatee gene of the pathway. As a result, the regulator genes of the network must be ranked according to their overall gain on *Gal2*. To do this, one needs to perform the same procedure conducted above to find the overall gain between *Gal4* and *Gal6*. The overall gains on *Gal2* from

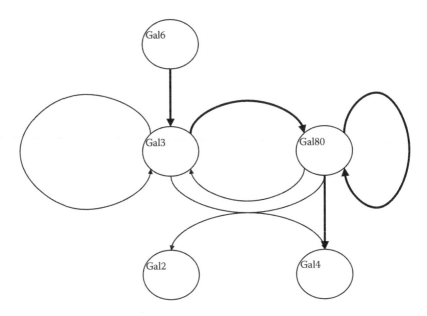

**FIGURE 9.8**   Path $P_4$ of obtained pathway.

*Gal3, Gal80,* and *Gal6* are calculated $(Gal2/Gal6)=-1187.7$, $(Gal2/Gal3)=0.16$, and $(Gal2/Gal80)=1119$.

As shown above, *Gal6* and *Gal80* have significantly stronger regulatory effect on *Gal2* compared with *Gal3*. Therefore, if *Gal6* had not been knocked out yet, the algorithm would have selected *Gal6* as the next gene to be knocked out. In this particular study, since the data set already contains the knockout experiments for all genes, the perturbation measurements were obtained after *Gal6* deletion data were already provided in the database. However; if more genes are added to the galactose pathway, the method systematically introduces other genes to be deleted in laboratory experiments.

## 9.6   SUMMARY

A method is described for inferring gene regulatory pathways using the integration of control theory, graph theory, and metabolic control theory. This method uses a correlation technique to obtain the most similar genes to the regulatee gene of the targeted pathway and then uses a reverse engineering method based on MCA to obtain the initial gene regulatory network. The method finally applies Mason's rule to find the overall interaction of each pair of genes in the network. Using Mason's rule, the most important regulators of the regulatee gene(s) of the pathway are recognized and introduced as the optimal genes to be knocked out in the next DNA microarray experiment. The method can be further generalized to incorporate the effect of external factors such as drugs.

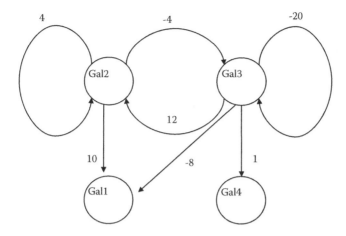

**FIGURE 9.9**    Pathway.

## 9.7    PROBLEMS

1. Consider the pathway given in Figure 9.9. Assuming that G1 is the main gene of the pathway, use the Mason rule to calculate:
   a. The overall gain from G4 to G1
   b. The overall gain from G3 to G1
   c. The overall gain from G2 to G1

2. Which one of the regulators, G2, G3, or G4, has the most effect on G1?

## REFERENCES

Darvish, A., and K. Najarian, "Prediction of Regulatory Pathways Using Metabolic Control and Mason Rule: Application to Identification of Galactose Regulatory Pathway." *Systems Biology* special issue, *Biosystems* 83 (February–March 2006): 125–135.

De la Fuente, A., P. Brazhnik, and P. Mendes, "Linking the Genes: Inferring Quantitative Gene Networks from Microarray Data." *Trends in Genetics*, 18 (2002): 18, 395–398.

Hofmeyr, J. H., and A. Cornish-Bowden, "Co-response Analysis: A New Experimental Strategy for Metabolic Control Analysis." *Journal of Theoretical Biology*, 182 (1996): 3, 371–380.

Hofmeyr, J. H. S., A. Cornish-Bowden, and J. M. Rohwer, "Taking Enzyme Kinetics Out of Control—Putting Control into Regulation." *European Journal of Biochemistry*, 212, (1993): 3, 833–837.

Ideker, T., V. Thorsson, J. A. Ranish, R. Christmas, J. Buhler, J. K. Eng, R. Bumgarner, D. R. Goodlett, R. Aebersold, and L. Hood, "Integrated Genomic and Proteomic Analyses of Systematically Perturbed Metabolic Network." *Science*, 292 (2001): 929–933.

Ito, T., T. Chiba, R. Ozawa, M. Yoshida, M. Hattori, and Y. Sakaki, "A Comprehensive Two-Hybrid Analysis to Explore the Yeast Protein Interactome." *Genetics*, 98 (2000): 8, 4569–4574.

Jensen, P. R., and K. Hammer, "Artificial Promoters for Metabolic Optimization." *Biotechnology and Bioengineering*, 58 (1998): 2–3, 191–195.

# 10  System Identification and Control Theory for Dynamic Modeling of Biological Pathways

## 10.1  INTRODUCTION AND OVERVIEW

In this chapter, the applications of system identification theory and nonlinear factor analysis in the processing of the time series of DNA microarrays are described.*

## 10.2  METHODOLOGY

Analyzing time series of molecular measurements, such as the DNA microarray time series, is a major step in addressing important problems in molecular biology. The gene expression is a temporal process, as the mRNA of each gene is affected by the mRNAs of other genes in previous time steps. Moreover, when a perturbation is applied to almost any biological system, the expression values of the majority of genes will be affected in the future times. Thus, to make an accurate analysis of the functional role(s) played by each gene, it is necessary to analyze time series DNA microarray measurements instead of static data. Unlike static microarray data, time series microarray contains the information regarding the dynamic correlations and interactions among genes that occur in time. However, as mentioned in most of the early works, in analyzing time series data, use the methods that have been designed and specialized for static data.

An important field of study that highly benefits dynamic models of gene networks is rational drug design. In many drug discovery applications, while knowing the steady-state effects of a drug is vital, the drug's short-term activities and potential transient effects on the molecular level must also be thoroughly studied and analyzed. Dynamic models of gene interactions, in addition to inferring dynamic gene regulatory network, allows predicting the trend of all genes in the future times, which in turn provides a predictive tool that can help the biologist simulate the drug's short- and long-term effects. Being able to at least predict the drug effects on the

* Portions reprinted, with permission, from A. Darvish, K. Najarian, D. H. Jeong, and W. Ribarsky, "System Identification and Nonlinear Factor Analysis for Discovery and Visualization of Dynamic Gene Regulatory Pathways," *Proceedings of the 2005 IEEE Symposium on Computational Intelligence in Bioinformatics and Computational Biology*, San Diego, California, November 2005. ©2005 IEEE.

genes involved in a biological system can significantly reduce the costly laboratory experiments involved in a typical drug design process. As a result, inferring dynamic gene regulatory networks that can be extracted from time series DNA microarray data has recently attracted a tremendous amount of attention. For instance, there has been a growing interest in using dynamic Bayesian networks to model gene expression time series.

In this chapter, we describe an algorithm that combines the capabilities of independent factor analysis with autoregressive exogenous (ARX) models to create a practical system for modeling of the interactions among all genes involved in a biological system. In a direct use of ARX models for gene network analysis, gene expressions can be considered as the output variables of the ARX model. The values of these variables for the future times can then be estimated using the model. There are some serious practical issues in the direct use of ARX for the modeling of the microarray time series. The main problem is the fact that in such an approach, considering the typical number of genes involved in a biological system, one would need an extremely large number of training points (in time) to reliably estimate all model coefficients. Because of several factors including the cost of conducting sample preparation and high-throughput experiments, many such time series have only a few time steps in them and, therefore, may not be sufficiently long to estimate a large number of model parameters. In addition, a blind application of ARX models to molecular biology problems would not provide insightful clustering interaction of the genes involved in a biological process, i.e., the model would fail to display the massive grouping and parallelism in genetic networks. A major difference between the method described in this chapter and other traditional applications of ARX models is the way the large number of variables (i.e., genes) is handled.

To address these issues, the algorithm described in this chapter exploits the fact that many genes behave rather similarly in a typical biological system, and, therefore, the role and effects of these genes can be somehow combined by a suitable clustering technique before dynamic modeling. *K*-means algorithm was used to cluster all genes to an optimal number of clusters and then applied the ARX algorithm to prototypes of clusters. This process significantly reduces the parameters of the ARX model. One disadvantage of this method is that it finds the dynamic model of the interactions among the prototypes of the genes classes and, therefore, may not predict the expression values of individual genes with high accuracy. The second issue is that *K*-means assumes a linear model for gene interaction, which as mentioned before, cannot reflect the nonlinear nature of gene networks. The assumption of linearity can dramatically limit the model accuracy.

In this chapter, the issue is addressed by replacing the *K*-means method with techniques based on independent factor analysis. We will describe models that utilize linear and nonlinear factor analysis methods together with ARX modeling. The main advantage of independent factor analysis is that it can extract the major factors of the data from the observed microarray time series. The method also allows reconstructing the gene network. Specifically, in this chapter, we show that if linear independent factor analysis is used, one can systematically reconstruct the gene regulatory network. To address the second issue, i.e., nonlinear nature of gene regulation, in

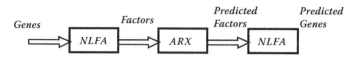

**FIGURE 10.1** Block diagram of algorithm.

this chapter, we describe the nonlinear version of the model that utilizes a nonlinear independent factor analysis (NLFA) and can effectively model the nonlinearity of microarray time series. To explore the advantages of nonlinear model over the linear models, both models have been used to predict the expression value of nonlinear synthetic gene expression data as well as a benchmark molecular data set containing the gene expression time series measured for the cell cycle process. The comparison of the results will indicate the capabilities and limitations of each approach.

The schematic diagram of the method is depicted in Figure 10.1. As shown in Figure 10.1, the method applies NLFA to extract the main trends (factors) of the data and then apply the ARX model directly on the obtained factors to predict the future values of the factors. The nonlinear factor analysis, as mentioned above, may be replaced by a linear factor analysis, which will result to a simpler overall model. For the reasons mentioned above, this method, unlike the typical system identification applications, applies the ARX modeling on the factors as opposed to the individual genes. Since the number of factors is significantly smaller than the number of observations (i.e., gene expressions), a significant reduction on the number of ARX model parameters is achieved. This reduction is in the order of $O(n^2)$, where $n$ is the number of inputs to the ARX model. The ARX model describes the interaction among all factors in two consecutive times, i.e., the value of each factor in time $t + 1$ is predicted from the value of all factors in time $t$. Then, as can be seen in the last block of Figure 10.1, the nonlinear combination of the factors at time $t + 1$ is utilized to predict the value of individual genes at time $t + 1$. This last step is performed simply by applying the inverse of NLFA as described later.

## 10.2.1 NONLINEAR FACTOR ANALYSIS

The nonlinear factor analysis is implemented using a backpropagation neural network that estimates the links between factors (sources) and observation, i.e.,

$$x(t) = f(y(t)) + n(t)$$

$$= C \tanh(Dy(t) + g) + h + n(t) \tag{10.1}$$

In Equation (10.1), $x(t)$ is the observation vector at time $t$ (e.g., gene expressions at time $t$), $y(t)$ is the vector of factors at time $t$, the nonlinear function $f(\cdot)$, which in this study is assumed to be tanh, is implemented as the activation function applied in neurons of the hidden layer, $C$ and $D$ are the weight matrix of the first and second layers, respectively, $g$ and $h$ are the biases of neurons in the hidden and output layers, respectively, and $n(t)$ represents the noise of the systems. Since in nonlinear factor analysis the sources (trends or factors) are assumed to have a Gaussian distribution, the posterior probability density of the unknown variables can be approximated by a Gaussian distribution, and the variances of the Gaussian distributions of the model are parameterized by the logarithm of the standard deviation, log SD, because then the posterior distribution of these parameters will be closer to Gaussian, which then agrees better with the assumption that the posterior is Gaussian. The noise $n(t)$ is assumed to be independent and Gaussian, therefore,

$$x(t) \sim N(f(y(t), e^{2v_x})) \tag{10.2}$$

Each component of the vector $v_x$ gives the log SD of the corresponding component of $x(t)$. The sources are assumed to have zero mean Gaussian distributions, and again, the variances are parameterized by log SD $v_y$.

$$y(t) \sim N(0, e^{2v_y}) \tag{10.3}$$

The main objective of ensemble learning is to approximate the posterior pdf of all unknown variables in the model. This means that from the observations $x(t)$, we need to estimate the statistical properties of all unknown values of the model including the sources and parameters, denoted as $\xi$. For this, we first define a cost function as the misfit between the actual posterior pdf $p(\zeta|x)$ and its approximation $q(\zeta|x)$. The posterior can be approximated as a product of independent Gaussian distributions:

$$q(\zeta|x) = \prod_i q(\zeta_i|x) \tag{10.4}$$

For each individual Gaussian distribution, we can use the posterior mean $\bar{\zeta}$ and variance $\tilde{\zeta}$. The cost function $C_\zeta(x; \bar{\zeta}, \tilde{\zeta})$ can be defined between actual posterior $p(\zeta|x)$ and its factorial approximation $q(\zeta|x)$. The optimization of cost function would be done with respect to posterior mean $\bar{\zeta}$ and posterior variance $\tilde{\zeta}$. The details of optimization technique have been described by Lappalainen et al. (2000).

After optimization, the estimates of all source (factor or trend) signals as well as parameters of the network are available. This allows us to feed the estimated factor signals to an ARX model that relates the present values of factors to their past values.

## 10.2.2 ARX MODEL

Once the major factors (trends) of the gene expression observations have been identified, an ARX model is applied to relate the expression levels of each of the factors to each other in time. Next, a brief description of the model is given.

The model relates the future expression level of each factor to the values of other factors in the past time(s). The model also incorporates the uncertainty in the model by considering an additive noise factor, i.e., $e(t)$ in the equations, the model is a linear system of difference equations, i.e.,

$$y_i(t) = -a_{i11} y_1(t-1) - \ldots - a_{i1n_1} y_1(t-n_1)$$

$$-a_{i21} y_2(t-1) - \ldots - a_{i2n_2} y_2(t-n_2)$$

...

$$-a_{ip1} y_p(t-1) - \ldots - a_{ipn_p} y_p(t-n_p) +$$

$$+ b_1 u(t-1) + \ldots + b_k u(t-n_k) + e(t) \tag{10.5a}$$

where $y_i$ is the value of the factor $i$, $p$ is the number of trends (factors), $n_j$ is the degree with respect to factor $j$, coefficient $a_{ipn_p}$ is the parameter relating the expression of the $i$th trend at time $t$ to the expression of the $p$th trend at time $t - n_p$, $u(t)$ is the exogenous drug or environmental input at time $t$, $n_k$ is the degree with respect to the exogenous input $u$, and $e(t)$ is the noise factor. For simplicity, for the rest of this section, we focus on the cases where no exogenous input is present. The extension of this formulation to the models with exogenous inputs is straightforward. Next, we define the matrix $y(t)$ as the matrix of all trends at time $t$, i.e.,

$$y(t) = [y_1(t), y_2(t), \cdots, y_p(t)] \tag{10.5b}$$

Also, define $\theta$ as the vector of all parameters, i.e.,

$$\theta = (a_{i11}, a_{i12}, \ldots, a_{i1n_1},$$

$$a_{i21}, a_{i22}, \ldots, a_{i2n_2}, \ldots,$$

$$a_{ip1}, a_{ip2}, \ldots, a_{ipn_p}) \tag{10.6}$$

To predict the coefficients of Equation (10.5), for each time step, the function $\varepsilon(t,\theta)$ is defined as the prediction error:

$$\varepsilon(t,\theta) = y(t) - \hat{y}(t|\theta) \tag{10.7}$$

where $\hat{y}(t,\theta)$ is the predicted output given the set of parameters $\theta$. The prediction error sequence can be processed through a stable linear filter $L(q)$ to further specialize the error function:

$$\varepsilon_F(t,\theta) = L(q)\varepsilon(t,\theta) \tag{10.8}$$

where $q$ stands for an element of delay. Next, we define $V_N(\theta, Z^N)$ as the measure of the total error (averaged over all $N$ time points):

$$V_N(\theta, Z^N) = \frac{1}{N} \sum_{t=1}^{N} l(\varepsilon_F(t,\theta)) \tag{10.9}$$

The function $l(\cdot)$ is any scalar-valued positive measure function (often defined as the square function). The parameter estimation is then defined as finding a set of parameters that minimizes the total error function, i.e.,

$$\hat{\theta}_N = \hat{\theta}_N(Z^N) = \underset{\theta \in D_M}{\arg\ \min} V_N(\theta, Z^N) \tag{10.10}$$

In this chapter, the square function is used as $l(\cdot)$, and the least squares method is applied to obtain the best set of parameters. Assume that:

$$\hat{y}(t|\theta) = \varphi^T(t)\theta \tag{10.11}$$

where $\varphi$ is the regression vector defined as:

$$\varphi(t) = [-y(t-1) \quad -y(t-2) \quad ... \quad -y(t-n)]^T \tag{10.12}$$

From Equation (10.11), the prediction error becomes:

$$\varepsilon(t,\theta) = y(t) - \varphi^T(t)\theta \tag{10.13}$$

Assuming $L(q) = 1$ (i.e., identify filter), and $l(\varepsilon) = (1/2)\varepsilon^2$ (i.e., the square function), the total averaged error criterion function becomes:

$$V_N(\theta, Z^N) = \frac{1}{N} \sum_{t=1}^{N} \frac{1}{2} [y(t) - \varphi^T(t)\theta]^2 \tag{10.14}$$

This is the least squares criterion for the linear regression, and it can be minimized analytically, which gives the following solution:

$$\hat{\theta}_N^{LS} = \arg\min V_N(\hat{\theta}_N, Z^N)$$

$$= \left[ \frac{1}{N} \sum_{t=1}^{N} \varphi(t)\varphi^T(t) \right]^{-1} \frac{1}{N} \sum_{t=1}^{N} \varphi(t)y(t) \tag{10.15}$$

Despite the detailed equations concerning the parameter optimization process, it can be seen that two major equations make the model:

$$x(t) = f(y(t)) + n(t)$$

$$= C\tanh(Dy(t) + g) + h + n(t) \tag{10.16a}$$

$$y(t) = A_1 y(t-1) + e(t) \tag{10.16b}$$

where Equation (10.16a) represents the NLFA and Equation (10.16b) describes the autoregressive (AR) model. $A_1$ is the coefficient matrix of the AR model.

### 10.2.3  LINEARIZED MODEL: EXTRACTING PATHWAY LINKS FROM NONLINEAR MODEL

It is often desirable, at least from the point of view of biologists, to form a "link-based" gene network that allows visualization of interactions among genes. The nonlinear model described above, although it is more accurate than linear models, cannot provide such more visually insightful link-based networks. In the literature of gene network estimation, direct links between genes are often considered as a more intuitive method of modeling and understanding the interaction among genes. Our objective is to obtain a linear approximation of the nonlinear model that allows directly relating gene expressions at time $t$ (and not just trends in time $t$) to the gene expressions in time $t + 1$. This can be done by manipulating Equation (10.16) together with a linear approximation as described below.

Form Equation (10.16):

$$y(t) = A_1^t \, y(0) + e(t) \tag{10.17}$$

Rewriting Equation (10.16.a) between times $t - 1$ and $t - 2$ and combining it with Equation (10.17):

$$x(t-1) = f(A_1^{t-1} y(0)) + n(t)$$
$$= C \tanh\left(D A_1^{t-1} \, y(0) + g\right) + h + n(t-1) \tag{10.18}$$

After linear approximation of tanh and ignoring the noise:

$$x(t-1) = C(D A_1^{t-1} \, y(0) + g) + h \tag{10.19}$$

Defining $G = A_1^{t-1} y(0)$, Equation (10.19) can be written as

$$x(t-1) = C(DG + g) + h \tag{10.20}$$

Thus,

$$G = K_{CD}\left(x(t-1) - h - Cg\right) \tag{10.21}$$

where $K_{CD}$ is pseudo-inverse of $CD$, i.e.,

$$K_{CD} = (CD)^T \left[(CD)^T (CD)\right]^{-1} \tag{10.22}$$

Now, Equation (10.1) can be approximated as

$$x(t) = C(DAG + g) + h + n(t) \tag{10.23}$$

Combining Equations (10.21) and (10.23),

$$x(t) = CDAK_{CD}\, x(t-1) + F \tag{10.24}$$

where $F$ stands for all terms that do not include $x$, defined as

$$F = -CDAK_{CD}h - CDAK_{CD}Cg + Cg + h \tag{10.25}$$

As it can be seen, Equation (10.24) provides a linear reconstructive model between the gene expressions at time $t$ and $t-1$ as

$$x(t) = A_{rec}\, x(t-1) + B_{rec} \tag{10.26a}$$

where

$$A_{rec} = CDAK_{CD} \tag{10.26b}$$

and

$$B_{rec} = F \tag{10.26c}$$

The model specified by Equation (10.26) and the parameter set $(A_{rec}, B_{rec})$ presents the direct linear dynamic relationship among genes, i.e., it provides the direct links between the gene expressions at time $t$ and time $t+1$.

## 10.3   MODELING OF CELL CYCLE: A CASE STUDY

The data set used to show the applicability of the above-mentioned method in real applications is the budding yeast *Saccharomyces cerevisiae* cell cycle data set introduced by Cho et al. (1998). In this study, the genes in the cell cycle data are clustered according to their known biological functions, e.g., the stage at which the genes are active. It is known that there are five major phases in cell cycle development: early G1 phase, late G1 phase, S phase, G2 phase, and M phase. The functional gene clusters are formed based on the activation of genes in one of the five phases, i.e., the genes in each cluster are the ones active in only one of the five stages of cell cycle. The data set used to create the clusters comprises the mRNA transcript levels of all studied genes during the cell cycle of the budding yeast *S. cerevisiae*. To obtain synchronous yeast culture, cdc28-13 cells were arrested in late G1, raising the temperature to 37°C, and the cell cycle was reinitiated by shifting cells to 25°C. Cells were collected at 17 time points taken at 10-min intervals, covering nearly two cell cycles.

The number of total genes considered in this study is 40. To train the model, the first 10 time points of each factor are used as training data to find the NLFA and ARX models. Then the expression values of each factor, and as a result, all individual genes, are predicted for all time steps. Since the number of time points in the training data is small (i.e., eight steps in each cycle), to minimize the number of parameters, the degree of the model is set to 1.

Following the above formulation, training of an AR model is equivalent to the estimation of $a_{ij}$ coefficients, where $i = 1,...,5$ and $j = 1,...,5$. Since the degree of the model is set to 1 for all genes, the third index of the parameters is dropped. The prediction results for two of the genes are shown in Figure 10.2. As can be seen in Figure 10.2, the predicted values match very well with the true expression values of genes indicating that the model can successfully predict the trend of the gene expressions based on the expression values of all genes at the previous times.

In the previous section, it was shown that for the synthetically generated nonlinear data, the nonlinear method has significant advantages over linear methods. Next, we show these advantages in the case of real DNA microarray. For the yeast S. cerevisiae cell cycle data, the expression value of the individual genes in the future times have been predicted using both linear and nonlinear methods. Figure 10.2 shows the predicted future expression using the nonlinear method as well as the actual expression signals for some important genes, and Figure 10.3 depicts the histogram of correct prediction for linear and nonlinear methods. As it can be seen from Figure 10.3, again, the histogram corresponding to the nonlinear method has a higher value for peaks in highest percentages compared with the histogram for the linear method. These histograms are further witness to the fact that, for a reliable and accurate estimation of gene networks, the nonlinear nature of the gene regulatory networks must be taken into consideration through the use of nonlinear models.

Next, we explore another important aspect of dynamic gene regulator networks, which is the stability issue.

## 10.3.1 STABILITY ANALYSIS OF SYSTEM

Because the degree of the model for each gene cluster was assumed to be 1, the AR model we developed for the cell cycle system describes the following discrete state space model:

$$Y(t) = A\ Y(t-1) + e(t) \qquad (10.27)$$

where $Y(t)$ is the vector of factors (trends) at time $t$. In the above model, the values of the factors play the role of the states in the state model. As in any other dynamic system, we are interested in assessment of the system stability. A straightforward approach in evaluating the stability in Lyapunov sense is examining the eigenvalues of matrix $A$, which are indeed the poles of the system. Every equilibrium state of the model described in Equation (10.27) is stable if and only if the magnitude of all eigenvalues of $A \leq 1$. If there is one or more eigenvalues whose magnitude equals 1, then the system oscillates. Although oscillation might be undesirable in many man-made systems, the survival of many biological systems (such as cell cycle) highly depends on it.

Figure 10.4 shows the poles of the system. As it can be seen in Figure 10.4, two poles of the system lie well inside unit circle, and the other two are almost on the unit circle. Having two poles on the unit circle describes why the cell cycle system is an oscillating and repetitive process. Since we obtained the AR equation using only 10 time points, there might be a small error in estimating the location of the

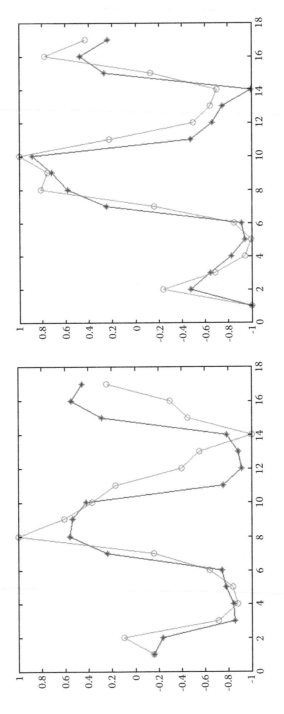

**FIGURE 10.2** Actual and estimated normalized expression values of two main genes: o = real values; * = estimated values.

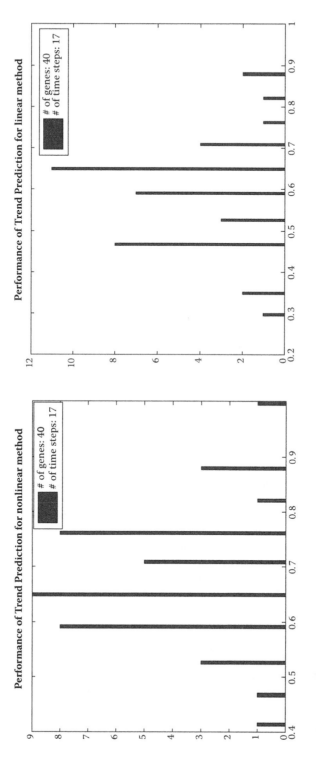

**FIGURE 10.3**   Histogram of correct trend predictions (cell cycle yeast data).

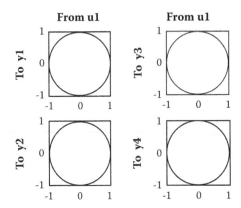

**FIGURE 10.4**  Poles diagram of ARX model.

poles. However, this result further witnesses the capability of the model in analyzing the biological systems under study. The stability analysis can be used to assess the system performance for many other gene regulatory pathways involved in important biological systems.

## 10.4  SUMMARY

In this chapter, we investigated analyzing the time series gene expression data. This chapter intended to address two problems: first, the reliable and accurate prediction of the gene expression values for the future time steps in a dynamic process. We described a robust prediction model based on ARX and NLFA techniques. This method neither makes any assumptions on the linearity for the underlying pathway nor attempts to oversimplify the model as in many other gene network estimation methods. The nonlinear factor analysis gives a significant ability in terms of dimension reduction before and after the ARX model is used. This reduction in dimensionality of the ARX model allows the use of this method for many short gene expression time series. The case study proved that the relationship between genes can be described far more accurately using nonlinear models as opposed to linear models. The second problem addressed in this chapter is reconstructing the dynamic regulatory network. The neural network–based structure of NLFA used in the technique provides the ability to reconstruct the observations (i.e., gene expressions) from the sources (i.e., trends or factors) even though the relation between the observations and trends is nonlinear.

## 10.5  PROBLEMS

1. What is the main disadvantage of the direct use of ARX model on DNA microarray time series without the use of nonlinear factor analysis?
2. For an oscillatory system, using a linearized model, what would be the location of the eigenvalues of the model?
3. Why do we expect that the cell cycle system will be an oscillatory system?

## REFERENCES

Cho, R. J., M. J. Campbell, E. A. Winzeler, L. Steinmetz, A. Conway, L. Wodicka, T. G. Wolfsberg, A. E. Gabrielian, D. Landsman, D. J. Lockhart, and R. W. Davis, "A Genome-Wide Transcriptional Analysis of the Mitotic Cell Cycle." *Molecular Cell*, 2 (1998): 65–73.

Lappalainen, H., and A. Henkda, "Bayesian nonlinear independent component analysis by multi-layer perceptions" in Girolami, M., Ed., *Advances in Independent Component Analysis*, Springer-Verlag, (2000): 93–121.

# 11 Gene Silencing for Systems Biology

## 11.1 INTRODUCTION AND OVERVIEW

Gene silencing is both a naturally occurring phenomenon and an experimental technique. When acting as the former, it helps gene regulatory networks maintain homeostasis, its biological status quo; when used as the latter, it is a tool to explore the gene regulatory network itself. This chapter begins with a review of gene silencing mechanisms and then proceeds to a review of the use of gene silencing as an exploratory method.

## 11.2 A BRIEF REVIEW OF RNA INTERFERENCE, SMALL INTERFERING RNA, AND GENE SILENCING MECHANISMS

Gene silencing is the process by which the expression level of a target gene is diminished. When it is done intentionally, this may also be referred to as "gene knockout" or "gene knockdown." As might be expected of a process as complicated as gene expression, there are many different ways in which genes can be silenced, principally diverging in the location and method of interfering with the normal conditions of the cell.

Recall that the central dogma describes the flow of genetic information within a cell from DNA to mRNA via transcription and then from mRNA to proteins at the ribosome via translation. Genes may be silenced both before and after transcription; although intentional gene silencing induced by external researchers is focused solely on posttranscriptional gene silencing for reasons that will be made clear in the next two sections.

### 11.2.1 PRETRANSCRIPTIONAL GENE SILENCING

Before a gene can be expressed, it must be accessible. Pretranscriptional gene silencing occurs when the DNA is made inaccessible or otherwise unsuitable for RNA transcriptase to perform its work.

Physical access is an obvious first opportunity for gene silencing. Histones are proteins that are built from eight recurring units roughly arranged to form a cube (two squares of four atop each other) around which the DNA molecule can wind itself. These repeating units of histones and their attendant DNA loops can be organized into chromatin fibers that allow for a compact form factor for the entire package. When they are densely packed, the individual genes on the DNA strand are not easily accessible to other chemicals within the cell and, hence, are unlikely to be expressed to any significant degree. Therefore, chemicals that can induce DNA and histones to form densely packed chromatin fibers are chemicals that will reduce the expression levels of genes. Unfortunately, this is not a particularly discriminating method of silencing genes. There is at least one way, presented in a section later within this chapter, by which histones specific to a target gene can be methylated and made to close ranks.

A second way to make the gene physically inaccessible is by using repressors. These are proteins that bind to noncoding regions of DNA near the promoter region, and their sole function is to interfere with RNA transcriptase binding to the promoter region. By doing so, they prevent the transcription process from beginning. An advantage of repressors is that they may be designed to act with some specificity; a clear disadvantage is—at least in eukaryotes—the problem of introducing them within the cell's nucleus.

Methylation of the DNA within the promoter region is a chemical way to decrease the expression level of the gene because it not only interferes with the binding of RNA transcriptase but can actually attract protein complexes that encourage the histones to form densely packed chromatin fibers, sequestering the promoter regions. Methylation is not a good way of silencing genes, due to the difficulties of targeting the specific promoter region of interest.

### 11.2.2  Posttranscriptional Gene Silencing

Once the gene has been transcribed to mRNA, the mRNA diffuses out toward the ribosomes where it will be translated into protein. Posttranscriptional gene silencing interferes with the mRNA molecules after they have been transcribed, preventing their translation. Again, there is more than one way to interfere with the mRNA: It can be physically altered so that it no longer can interact with the ribosome and it can be degraded before it reaches the ribosome.

One of the conceptually simplest ways to alter mRNA so that it cannot interact with the ribosome is by introducing antisense RNA. This is a single strand of RNA that is complementary to some portion of the mRNA produced by the target gene. Once the antisense RNA binds to the mRNA molecule, the larger complex no longer is suitable for processing within the ribosome, and no resulting protein is produced. Antisense RNA can be built to target specific genes, but it has a few drawbacks, one of which is the difficulty in delivering the product to the desired location. Morpholinos—morpholino-modified oligonucleotides—are antisense molecules that can bind to mRNA but use a different backbone chemistry; they have proven to be effective in conducting gene silencing experiments in a wide array of organisms.

Degrading the mRNA, although it may sound drastic, is of course another effective mechanism of silencing a gene after transcription. Conveniently, there happen to be complexes within cells that are very effective at targeting and destroying mRNA fragments. The sole challenge is to deliver the triggering signals effectively. The most promising posttranscriptional gene silencing mechanism for researchers, and one that operates principally by breaking down the mRNA, is RNA interference (RNAi).

### 11.2.3  RNA Interference

*RNAi* is a process of gene silencing that may have evolved as a part of an immune response to viral DNA, although it has other uses in downregulating genes. Described by Fire and Mello in 1998, RNAi occurs when small interfering RNA (siRNA) molecules that are complementary to the target mRNA become associated with an RNA silencing complex (RISC). This complex, incorporating the siRNA guide strand, can then bind to the target mRNA, at which point an endonuclease, argonaute, is activated, which destroys the mRNA. (There is also a second pathway, in which the siRNA binds to a RNA transcriptional silencing complex that can target histones or methylate DNA at the target gene site.)

One of the keys to RNAi is that the guide siRNA fragments are relatively short, on the order of 21 nucleotides long. This is enough information to allow for specificity in selecting a target but also prevents legitimate mRNA from being used itself as a guide strand.

The source of the RNA strands that guide the RISC to mRNA targets can be both endogenous (produced within the cell) and exogenous (produced outside the cell).

In the endogenous case, DNA that codes for micro-RNA (miRNA) is expressed, and these small fragments eventually can bind with RISC just as the siRNA fragments do. This mechanism is useful for the self-regulation of genes. The DNA that codes for the miRNA should be preferentially expressed when there is too much of a particular protein; the miRNA would then interfere with the mRNA, which would result in more of that overabundant protein being created, helping to maintain the cell's internal balance.

In the exogenous case, siRNA from viruses or researchers is used to silence target genes. There is some evidence that viruses—always in an arms race with their host cells—have evolved siRNA strands specifically to help disable key cellular defense mechanisms via RNAi. Researchers and drug designers, on the other hand, design and deploy their own siRNA fragments to perform knockout or knockdown analyses of the gene networks they study.

## 11.3  PATHWAY PERTURBATION USING GENE SILENCING

Genome sequencing efforts have produced a large amount of data, but each sequence is merely a static description of a genome that may have different expression levels at different times (and, in the case of multicellular organisms, different locations). High-throughput assays allow us to capture information about the expression levels of genes over time. These data help to provide the basis for inference of gene networks. They do not necessarily, however, inform us as to the purpose of these gene networks. Overexpression and gene silencing are complementary methods for exploring the relationships implicit in gene networks. When combined with computer modeling and simulation techniques, these two methods can help provide the data necessary for a systems biology approach to understanding the impact of gene regulatory networks on the single-cell and the (potentially) larger organism.

### 11.3.1  OVEREXPRESSION

In a healthy cell, expression levels of genes are (self-)regulated by networks of proteins (gene products) that interact with the environment. Proteins to add in the digestion of lactose, for example, may be repressed in the presence of glucose but may be upregulated when glucose is scarce, allowing for the metabolism of a secondary energy source. In this way, both underabundance and overabundance are adjusted so that the cell stays in homeostasis.

By either inserting additional copies of the target gene—by modifying the organisms' DNA or more simply by adding a plasmid—or introducing promoters, the total amount of a gene's expression can be increased beyond what would occur naturally. When the gene product is a protein that serves as nothing more than an inert building block for other assemblies, the overexpression may not cause any discernable change. That is, the phenotype may be unaffected by the overexpression of the target gene.

Because gene products can also influence other genes—both directly, by promoting and repressing the expression of genes, and indirectly, by altering the concentration of other strains in the cell's environment that in turn can influence other genes—the distention of gene products often has unexpected consequences.

By using high-throughput mRNA detection, the change in effect on all of the genes in the network can be assayed. When these network expression levels are coupled with observations of the resulting phenotypes, researchers gain additional insight into the relationships among the participants, especially when these overexpression data are combined with data from gene silencing and mutation experiments.

### 11.3.2  GENE SILENCING

Gene silencing is another effective research method for reverse engineering the genome: Instead of using phenotype to inform a search for genes, it silences interesting genes to identify phenotypic effects. In contrast with overexpression, gene silencing essentially "comments out" a gene in the network, removing its expression products from the cellular environment.

In concept, this is analogous to trying to understand how your car works by randomly cutting some of the belts or hoses. If the cell explodes, then this gene silencing is marked "lethal" and is assumed to be critical in the network under investigation. Multiple rounds of knockdown testing will accumulate successively more data that can fuel these inferences.

### 11.3.3  MODEL IMPROVEMENT AND VALIDATION

Clearly, gene silencing and gene amplification are approaches that derive additional benefit from being used together: Knowing the effects of both increasing and removing gene products will provide more information than either one half or the other will independently. If, for example, overexpressing a gene has no noticeable effect, but silencing the gene is lethal, then it may suggest that one of the gene products has a minimum threshold effect that is necessary for cell life; alternately, the absence of one of the gene products may in fact serve as a signal for cell death.

Considering that a single gene can, through alternate splicing, encode many different proteins, each of which may participate in the gene regulatory network independently, complicates the analysis of both the overexpression and silencing data considerably. Fortunately, this is the kind of problem that is suitable for modeling and simulation within a computer, and there are software packages that will assist with correlating observed expression levels and effects.

## 11.4 SUMMARY

Gene regulatory networks have evolved over billions of years and can be immensely complicated. The internal relationships among gene products can be sensitive yet remarkably tolerant of changing environments. The resilience with which cells can identify viral infection and take steps to combat rogue RNA are nothing short of remarkable. Yet, this same richness confutes our disentangling of the influences.

Gene silencing is of interest not only to scientists who wish to understand more about mapping the operational life of cells but is also of tremendous importance to those who wish to alter the same cells: from agronomists who wish to engineer more robust or disease-resistant crops, to medical researchers who wish to diagnose and treat cancers and other genetic diseases, to pharmaceutical companies who wish to take advantage of these networks to provide more focused drugs with fewer side effects. A thorough understanding of both natural and artificial gene silencing could have far-reaching impact.

## 11.5 PROBLEMS

1. What are the functional differences between siRNA and miRNA?
2. What software packages are used to help map gene regulatory networks?
3. Why must siRNA fragments in mammals remain relatively short?
4. What is a plasmid and how does it work to introduce genes into a cell?
5. What is the history and derivation of morpholinos as an antisense agent?
6. Find examples of organisms in which both overexpression and gene silencing have been used to elucidate some portion of a gene regulatory network.

## REFERENCES

Fire A., S. Q. Xu, M.K. Montgomery, S. A. Kostas, S. E. Driver, and C. C. Mello. "Potent and specific genetic interference by double-stranded RNA in *Caenorhabditis elegans*." *Nature* 391 (1998): 806–811.

# 12 Simulation and Systems Biology

The yellow fog that rubs its back upon the window-panes,
The yellow smoke that rubs its muzzle on the window-panes
Licked its tongue into the corners of the evening,
Lingered upon the pools that stand in drains,
Let fall upon its back the soot that falls from chimneys,
Slipped by the terrace, made a sudden leap,
And seeing that it was a soft October night,
Curled once about the house, and fell asleep.

**from "The Love Song of J. Alfred Prufrock" by T. S. Elliot**

## 12.1   INTRODUCTION AND OVERVIEW

The words in T. S. Elliot's *Prufrock* somehow make the miasma come alive and
seem real, not merely to the mind's eye but also to its fingers, its nose, its ears, its
feet. These eight lines capture and inject directly into the reader some core essence
of the fog and smoke character.

Simulation, when done well, ought to have the same kind of expressive power
using the same economy of form as poetry, whether it concerns the core human
experience or systems biology. In this chapter, we will review the nature (and challenges) of simulation and see how this applies to two in-depth case studies of simulation within systems biology.

## 12.2   WHAT IS SIMULATION?

Simulation, arguably, is the art of useful exaggeration: To simulate is to excise a
single theme from a larger complex, translate it into a smaller, simpler setting—often
distention, at least one dimension—and still have it evince the same characteristic
behaviors that it exhibited within its original context. Simulation blends both the arts
and the sciences even though our current perspective is science-centric, as Plato's
Socrates in *Republic* remarks:

> Now, when all these studies reach the point of inter-communion and connection with one
> another, and come to be considered in their mutual affinities, then, I think, but not till then,
> will the pursuit of them have a value for our objects; otherwise there is no profit in them.

These "mutual affinities" influence our consideration of both the power over expression and the economy of form.

### 12.2.1 Power of Expression

The power of expression, the ability to engage an audience and persuasively convey an idea, an image, an emotion, is easily attributed to the arts. Consider the impact, the sensory response that results from each of these:

- Matisse's *Blue Nude*
- Dilbert, Wally, and their PHB
- Michelangelo's David
- "Four score and seven years ago..."
- Châteauneuf-du-Pape
- "I know what you're thinking. *Did he fire six shots or only five?*"
- The beginning of *Thus Spake Zarathustra* as used within *2001 A Space Odyssey*
- "Less filling. Tastes great."
- "The tintinabulation of the bells, bells, bells..."
- The Taj Mahal

But what of the sciences? A 40-GB log file from a series of simulation runs does not necessarily excite the same emotional response as those examples. The expressive power of a simulation is not in the raw data output but instead requires translation and interpretation to become vibrant. This is similar to MP3 files: Nobody reads the raw bytes in the file on the hard drive, but we can all enjoy listening to the exact same sequence of bytes once it has been run through the interpreters and speaker systems; only then is the expressive power of the music set free. Simulation works in much the same way: A model that elegantly describes immigration and emigration patterns among Middle East countries might generate huge volumes of purely numeric output that nobody can interpret, but once there is a suitable interface in place, the patterns and relationships and ideas that the model embodies are set free.

The core challenge of designing useful simulations, then, is twofold:

1. The model must embody one or more meaningful relationships.
2. There must be some mechanism in place—an interpreter of some sort, be it visual or narrative or some other method—to allow the user to experience the results effectively.

This is the same challenge as in other artistic endeavors: First, decide what you want to say; second, decide how to say it so that the audience has the best possible exposure to the message. Or, as an engineer might express the same concept: Form follows function.

### 12.2.2 Economy of Form

Science and engineering have long been the bastion of economy. Occam's razor, although the name is relatively recent, has been one of the key principles of good design for the millennia. It was possible, pre-Copernicus, to generate maps that would correctly predict the position of major celestial bodies. They were correct inasmuch as they worked. They were dropped, however, when a simpler explanation (model) became available. Einstein expressed it this way:

> The supreme goal of all theory is to make the irreducible basic elements as simple and as few as possible without having to surrender the adequate representation of a single datum of experience.

This is oft paraphrased as: A model should be as small as possible, but no smaller. In the arts, there is a similar motivation. Edgar Allan Poe, in *The Poetic Principle*, argued against (over-)lengthy verse:

> It is to be hoped that common sense, in the time to come, will prefer deciding upon a work of Art rather by the impression it makes—by the effect it produces—than by the time it took to impress the effect, or by the amount of "sustained effort" which had been found necessary in effecting the impression.

In music, a leitmotif often consists of just a few measures. Within this small window, a composer can express core attributes of the character, mood, or action. A caricature captures in a few pen strokes the details that allow the viewer to identify by key attributes the identity and the character of the subject. The drama behind Lincoln's Gettysburg Address was not merely the subject matter but the fact that he was succinct, and yet, how much more powerful is the piece for its directness?

A simulation of Charlotte, North Carolina, might be more accurate if it includes a list of all of the books on my shelf, but it is not likely to be more useful because of those details. As in other arts, one of the key challenges to designing a good simulation is knowing which elements contribute to the main idea and having the courage to prune those that do not.

## 12.3   CHALLENGES TO EFFECTIVE SIMULATION

As described in the preceding sections, there are a number of challenges to creating effective simulations. Some of the most significant that you will have to address are, in order,

- Identifying meaningful themes
- Selecting details to use and exclude
- Validation

### 12.3.1   IDENTIFYING MEANINGFUL THEMES

Akin to asking, "Is a zebra black with white stripes or white with black stripes?" a simulation must begin with the identification of a core theme: What key property or behavior will be excised from the complex system so that it can be reproduced in caricature in the smaller test system? If you are studying epigenetics, then the core theme might involve the way that a relatively small handful of promoters, repressors, and genes can interact in unexpectedly complex ways to create rich patterns of gene expression; for population biology, it might involve modeling sexual selection and its impact on both phenotype and genotype across generations; for gene activation pathway analysis, the focus of the simulation might be diffusion of key enzymes or signals.

Clearly, the selection of a theme is a time-consuming process that requires a significant amount of background knowledge concerning the original system being modeled. It is also unrealistic to expect, even when you have managed to identify the important theme, that your first implementation or model will encompass the key characteristics fully, in part because of the importance that detail selection and validation play: Many rounds of modeling are generally required before adequate simulations result.

Selecting a theme is analogous to deciding what you wish to communicate or express: What is the thesis or main idea? If the reader, consumer, or reviewer is to understand only one idea from this work, what should it be? Answering this question can be one of the most difficult challenges to a simulation project.

### 12.3.2 DETAIL SELECTION AND REJECTION

As suggested by the Einstein quotation earlier, the construction of a useful simulation is like playing a theory-based game of limbo: How much detail can we discard before the core theme is compromised? In fact, the risks come from two directions:

1. Removing key details robs the simulation of verisimilitude, that is, it stops representing the important phenomenon in any meaningful way.
2. Allowing unnecessary details to persist in the simulation is bloat: Their presence suggests that they are important to the central theme, even though they may not contribute meaningfully to its development.

The appropriate level of detail may be impossible to identify *a priori,* because

- The impact of the choice may not be immediately obvious: What might it mean, when simulating protein interactions within a given environment, to assume that (not) all possible reactions occur immediately? Often, the only good way to resolve these questions is to try variants of the core simulation under both choices and compare results. Unfortunately, this approach becomes combinatorially intractable.
- Not all details are used immediately: A simulation of how proteins fold may work quite well on small proteins but may fail to produce adequate results as soon as the number of residues climbs above the assumed threshold for the model, exposing shortcomings in the level of detail that were not obvious when simulating smaller chains.
- Many simulations, if they are going to be interesting or useful, result in nonlinear dynamics, meaning that the simulations themselves often exhibit a sensitive dependence on initial conditions: This not only affects the specific data values used during initialization but also the properties that are (not) used as part of the model.

### 12.3.3 Validation

Validation is the task of assessing how useful your simulation is—how high its quality is—often using a quantitative metric such as "RMSD of the $C^\alpha < 5$ Å" or some other correlation of simulated outputs with real-world (observed) outputs. As discussed in the following sections, though, the comparisons are often not as simple as might be imagined: either because the two series to compare are not well-defined or because the simulations themselves are stochastic and require some sort of ensemble construction before comparisons to estimate validity can be computed.

#### 12.3.3.1 Measuring Quality Once

Consider a simple example, such as linear regression modeling. While this may not seem like much of a simulation, this type of statistical analysis agrees with our core definition: It extrapolates a pattern or behavior from a more complicated environment and introduces a simpler reduction. One thousand data points may be reduced to a single, straight line on a chart. Reading this line allows us to infer what might happen at any point not sampled. Even for this relatively simple case, though, there are many different ways to estimate the goodness of fit (the solution quality, or simulation validity), including mean squared error, $R^2$, and absolute error.

As a simulation becomes more complicated than linear regression modeling, the number of candidates for fitness functions (quality estimators) increases proportionately. It may be that the difference between the simulated output and the real, observed values are not equally important across all cases: If a gene is actually expressed 0%, it may be that a simulated expression level of 50% is still more desirable than a simulated expression level of −40%. Or, it may be that the expression level of some genes in a network simulation is more tightly controlled than others, allowing for a very fluid fitness function. The complexity in estimating fitness also increases as the number of simulated outputs increases: Simulated protein folding produces a single 3-D structure that contains many properties that can be validated: locations of the alpha carbon atoms, distribution of electrostatic charge, isolation of hydrophobic residues, patterns of side-chain packing, etc. Every single output property is a candidate for estimating the validity of the simulation; the only good way to select among all of these candidates is to first select a sufficiently narrow core theme from which the validation will flow naturally. This is another reason why simulations tend to be developed incrementally because the core theme to be simulated and the validation process may influence each other.

At the extreme end of the continuum are simulations that are not explicitly based on phenomena that are directly observable: Conway's Game of Life, for example, exhibits properties that might identify as "lifelike" but in a way that defies easy comparison with any specifically meaningful analog. Life is a simulation of a class of patterns, and these simulations are extraordinarily difficult to validate except within the mind of the observer.

#### 12.3.3.2 Handling Stochastic Results

For as difficult as it can be to estimate quality of a single set of simulation outputs, many simulations are stochastic in nature, that is, they rely on pseudorandom number generators within the computer as part of their design. This means that the same simulation can produce different outputs from the same inputs on two different runs. Now, validation becomes more complicated because instead of a single set of concrete outputs, there is a collection of disparate outputs to compare to the real-world data. In many problems within the life sciences, even the real-world data may not be crisp: Protein conformations are ensemble models, which is to say that they are somehow averaged across many different samples.

There are a few obvious ways to handle these uncertainties:

- Straight averaging: Ignoring the fact that both the real-world data and the simulated outputs are inconsistent, average the results in both cases and simply compute the difference of the means as your estimate of solution quality. Clearly, this carries with it inherent risks of overgeneralizing based on the degree of variation implicit within each average.
- Compare the distributions: Knowing the mean and the variance for each of the points that constitutes your samples may allow you to compare the distributions of results using the appropriate statistical test ($X^2$, Student $t$ test, etc.). This may also be a bit simplistic, depending upon your need.
- Temporal progression: If your simulation is time-based (and many are), then you may wish to abandon single-point ensembles in favor of across-time progression instances. This does not eliminate the group nature of your data but changes the basis for each data point in the sample and may be a way to reduce some of the variation.
- Weighted combination of criteria: Typically, you have multiple criteria you wish to compare, so they may be rolled into a single function wherein each term is weighted. Weights commonly are linear or exponential.

## 12.4  CASE STUDIES

We present two case studies here, the contrast of which highlights some important differences in domain, focus, and intent: Rosetta@Home and simulated protein computing (SPC). The first is a computer simulation of how real biology works; the second is a simulation showing how real biological systems might be used to perform future computing tasks. For each, we present

- An overview
- A brief description of its underpinnings and how it works
- Details that the designers chose to include and discard
- The source of the computing power
- Future directions

### 12.4.1  Simulated Protein Folding: Rosetta@Home

Rosetta@Home is a massively distributed version of a subset of the Rosetta Software Suite for simulating how proteins fold into their tertiary structure (3-D shape) given only their primary structure (sequence of amino acids). This topic was discussed at length in an earlier chapter within this text. The original Rosetta core software was designed at the Baker Laboratory within the University of Washington Department of Biochemistry. Rosetta@Home uses volunteer computing cycles from around the world to search for folded protein conformations that will eventually help treat human diseases such as Alzheimer's disease, HIV, and prostate cancer.

### 12.4.1.1  Generally How It Works

The protein is initially constructed from free amino acids by the ribosome. As it emerges, it begins to fold, interacting not only with itself but with the solvent and (possibly) helper proteins around it. Rosetta@Home is an ab initio model of protein folding, meaning that it assumes the protein folds in isolation. The model makes the assumption that the local in-protein residue effects predispose the protein to fold into select secondary structures (alpha helixes, beta sheets, etc.) and that the weaker nonlocal influences of the protein on its folded shape act by selecting the combination of secondary structures that represents the lowest energy tertiary conformation.

There are, hence, a few core steps to the Rosetta@Home protein-structure prediction algorithm, modeled as independent processes:

- Generate a collection of all 9-mers (sequences of 9 amino acid residues), and search the protein databases for known secondary structures in which each 9-mer participates. This defines a distribution of secondary structures that are possible at each subsequence. In one version of Rosetta, Bayesian mechanics were used to help compute the most probable secondary structures.
- Simulated annealing is used to find the low-energy conformations that are consistent between selected secondary structures and the nonlocal effects. Borrowed from metallurgy, simulated annealing is a search algorithm in which the temperature (energy, change) of the system from one state to the next is varied over time, allowing the solution to change more at the beginning of the search than at the end; in this way, it helps to fight the tendency of gradient-descent algorithms to settle within suboptimal local minima.

There are, of course, additional details that must complicate these methods: the initial generation of secondary structures must take into account nonviable combinations (that would see overlapping alpha carbons, for example), as well as nonstandard bond angles or clashing R groups. The simulated annealing itself is also a tricky method to calibrate; is it reasonable to expect a single set of optimization parameters to apply to proteins of arbitrary length?

### 12.4.1.2  Details to Use or Discard

As described in the preceding section, the environment in which proteins are constructed surely influences their final, folded conformation, and yet, it is not feasible to construct a simulation environment that models all of these species with complete fidelity, so the challenge to the modeler is to identify those aspects of the protein-folding process that are most likely to contribute to high-quality structure predictions, while at the same time eliminating all details that are unlikely to detract significantly from the final model quality.

Rosetta relies upon a two-part paradigm that is entirely inwardly focused. That is, ab initio protein folding is a modeling method that assumes that only interactions among the protein's own residues influence the final conformation. Unlike some models that represent water (surrounding solvent) as a constant background field property—or BlueGene/L that models the water molecules independently—Rosetta ignores the solvent entirely.

In contrast with many other approaches, Rosetta does not rely wholly upon a molecular dynamics (MD) approach to protein folding. The first phase, in which the secondary structures are inferred from a statistical analysis of common $n$-mers between the target protein and other proteins within public databases, serves as a proxy for these MD results. The implication is that similar (very short) sequences of amino acids will share MD properties and that the distribution of secondary structures found among other proteins will represent a good operational version of the likely conformations. Note that this is not a homology approach but rather a statistical approach.

The nonlocal effects that the Rosetta@Home Web site mentions include "hydrophobic burial, electrostatics, main-chain hydrogen bonding, and excluded volume." These are all common-sense, high-level dynamics for protein folding: Hydrophobic residues tend to be sequestered within the center of the protein, and hydrophilic residues tend to reside on the outside shell; electrostatic repulsion will influence the energy level of the possible folded conformations in much the opposite way that main-chain hydrogen bonding can strengthen different conformations; the excluded volume is an implicit recognition of the role the surrounding solvent plays in determining the final, folded conformation of the protein.

### 12.4.1.3  Computing Power

Rosetta@Home is built on the Berkeley Open Infrastructure for Network Computing (BOINC) platform developed by the University of California at Berkeley. This is the same platform that is used to power the SETI@Home and ClimatePrediction.net projects, among others. The design principles are simple:

- Large problems that are susceptible to supercomputer analysis are separable. That is, they can be broken down into smaller subproblems that can be solved independently. A fluid-flow simulation can be separated into separate, smaller volumes that can be analyzed independently. A protein folder can generate thousands of possible conformations that can be scored independently.
- The number of dedicated supercomputers in the world is much smaller than the number of small processors sitting idle around the world. Although estimates vary, even in 1995 (Anderson et al., 1995), it was estimated that 60% of workstations were 100% available during regular work hours; add this to the off-hours availability and this represents a significant amount of processing power that can be harvested. The SETI@Home project (as reported by http://boincstats.com/stats/project_graph.php?pr=sah) is averaging over 530 TeraFLOPS (billions of floating-point operations per second).
- Combining the two previous points, it is possible for a relatively small number of central project servers to distribute work across a network of

volunteer computing nodes (machines) distributed across the globe. As the clients complete computations on their small subproblems, they report the results back to the central servers that organize the results* and assign the next, separable portions of subproblems back out.

- Communication overhead is always a drag on computational efficiency, but improvements in network connectivity as well as the amount of local storage space—both RAM and hard-disk space—have improved through-put significantly.

Both businesses and private individuals volunteer their machines to these efforts. Rosetta@Home is hosted as a BOINC project, meaning that volunteers can specify that they wish their idle CPU cycles to contribute to this project's computing resources. This means that Rosetta@Home is effectively always running, globally distributed across a very large number of processors churning away for little cost to the project except maintaining the central project servers.

### 12.4.1.4 Future Goals

Although curing human disease is one of the significant long-term goals of the Rosetta@Home project, the short term is consumed with trying to understand how best to solve the problem of predicting tertiary protein structure from primary structure. Currently, inferring the tertiary structure requires crystallographic diffraction studies that are expensive in terms of time, labor, and materials. A reliable analytic tool that is faster and less costly would generate a significantly greater volume of 3-D structural data that would allow researchers to study:

- The docking and binding sites that involve residues that are spatially distributed across the linear sequence of amino acids but because close neighbors in the final, folded conformation. Hydrophobic residues generally end up in the interior of the folded protein, with the hydrophilic residues on the exterior shell, but the specifics of how the protein backbone folds has a dramatic influence on the final spatial correlation of alpha carbons along the chain.
- A better understanding of protein active sites also allows us to identify both new targets for existing pharmaceuticals and new opportunities for drug-based intervention. Viagra was first designed as a treatment for high blood pressure, but has been realigned with higher-margin markets in light of its alternate applications; similarly, as our understanding of protein function increases, so do our opportunities to spot points of failure where an effective pharmaceutical intervention might be able to interrupt a disease pathway.
- Effective knowledge about binding sites also allows pharmaceutical and medical researchers to design potential inhibitors for these proteins. This becomes a crucial step in the treatment of many diseases. The Rosetta@

---

* Including validation, typically, BOINC projects require that at least two copies of the same subproblem result agree before accepting them as final results. This is to prevent the poisoning of the results pool by malicious workstations.

Home site discusses prostate cancer, for example, in which the treatment goals are to prevent aggressive androgen receptors from either entering the cell nucleus or binding to DNA to express tumorgenic agents. To accomplish this will require the development of highly specialized inhibitors.

Although having an enumerated list of binding sites gives us better opportunities toward so-called rational drug design, the systems biology approach clearly provides benefit. A drug that may be designed to be effective for a single target disease protein may have deleterious effects when it is introduced elsewhere in the proteome. A systemic understanding of the proteins, and their folded conformations, becomes crucial in designing effective treatments.

To volunteer your spare CPU cycles to Rosetta@Home, please visit http://boinc. bakerlab.org/rosetta/rah_intro.php.

### 12.4.2 Simulated Protein Computing

*SPC* is a hybrid system designed to be (in the long term) both a simulation of how proteins interact and a general-purpose computing platform that relies upon a nondeterministic simulation of protein interactions to perform its work. Conceived and implemented at the University of North Carolina at Charlotte (and to be released under a GNU Public License version 3 in the fall of 2008), SPC is an effort to build an alternate computing paradigm to traditional von Neumann processing. While still in its early stages, SPC has already demonstrated that it can serve as a universal computing platform.

#### 12.4.2.1 Generally How It Works

In a traditional von Neumann fetch-execute architecture—the basis for almost all of the computing devices used today—instructions are fetched from memory one at a time, executed by the processor, and the results stored back into memory; this is an innately serial process that has not been substantially changed by advances in either processor design or compiler technology in the more than 60 years since von Neumann described it. Office workers know very well that a word processor runs at essentially the same speed on a single-processor machine as it does on a dual-processor or quad-core machine; this is because each program—a sequence of instructions—can be run through only one processing core at a time. High-performance computing is more about how to decompose monolithic computing problems into separable subproblems that can be farmed out to multiple processors in parallel than it is how to change the nature of computation.

SPC, as the name suggests, is an attempt to harness the parallelism inherent in protein-protein interactions within living cells. Furthermore, it seeks to take advantage of the fact that numbers in digital computing are analogous in many ways to proteins within a cell, serving both as structures and instructions. Instead of programming "$x \leftarrow 1 + a$," SPC would have an enzyme (functional protein, defined by the user) that could bind to a "1" and an "a" protein (data) and fuse them into a new protein tagged with an "x" marker. Instead of source code libraries, then, programmers prepare infusions of protein solutions. Although this may not seem significant, the impact of this change of focus has interesting implications.

These are SPC's core principles:

- Biological systems are time-dependent but operate with massively innate parallelism. Although there may be only a single copy of DNA within a cell's nucleus, there are myriad active portions being transcribed all at once; these result in many bits of mRNA that trigger the production of proteins that serve both structural and functional purposes.
- Real biological systems distribute proteins and cell structure nonuniformly: The molecules inside the cell nucleus are differently concentrated than those in the cytoplasm. Gradients are the motive force within this environment. Semipermeable membranes are one of the key (programmatic) classes of inhibitors to these gradients.
- Whether two molecules will interact is more than simply whether they are colocated but also depends on other factors such as orientation, competition, and environment (temperature, pH, etc.).
- The output from life sciences experiments is rarely a single observation but instead is much more likely to be some sort of assay or sample. The outcome, then, is more often than not inferred from a distribution of products.
- Although the von Neumann architecture has its bottlenecks, it is a universal Turing machine, meaning that it can compute all of the results of—and can, in fact, simulate—any other universal Turing machine. This is a useful property to have.

### 12.4.2.1 Details to Use or Discard

Encoding is the first challenge to consider, and it invites discussions both about what details are preserved and what details are discarded. The first issue is storage: In a digital computer, there can be only as many different variables as there are locations in memory; in a cell, the only limit on the number of different proteins in existence is conservation of mass. When simulating proteins as computing agents, this mismatch in storage is a problem: It is intractable for a digital computer—in which the proteins computing is being simulated, after all—to store all of these species that could reasonably be expected to exist within a cell: A polypeptide chain of only 50 residues could assume $20^{50}$ ($\sim 1.1 \times 10^{65}$) different combinations by itself.

The solution used within SPC is to allow for lossy (probabilistic) data storage. Borrowed from compression studies, "lossless" refers to a method that does not discard any data but can completely recreate the original (source) signal; "lossy" is the complement or a method that generally reproduces the important parts of the source signal but does not guarantee to reproduce the exact source signal. Figure 12.1 is an example of how SPC stores proteins: The given protein is recognized to have an implicit beginning-of-sequence (BOS) marker at the start and a corresponding end-of-sequence marker (EOS) at the tail. The entire sequence is then decomposed into a series of $n$-grams or shorter subsequences of uniform length. These $n$-grams are then collected into a single-state transition diagram (akin to a Markov model). The depth of the tree (the side of the $n$ in the $n$-grams) influences both the maximum storage required as well as the fidelity with which sequences can be recalled. In the example in Figure 12.1., $n = 3$ and the probabilistic transition tree happens to capture the

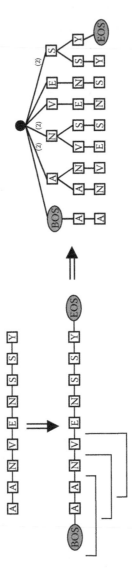

**FIGURE 12.1**   Encoding a small protein fragment in a 3-gram tree.

given sequence with perfect fidelity because starting with "BOS" necessarily entails "BOS-A-A," whose only choice for a follow-up residue (using the tree as a look-up) is "N" from the "A-A-N" branch. Following these reconstruction steps will reproduce the entire fragment in its entirety.

What this encoding preserves is the importance of local associations of amino acids. What this encoding discards is fidelity: In many cases, the proteins that are stored in these probabilistic transition trees cannot be reproduced perfectly, even though all of the $n$-grams within them are valid from the initial sample.

SPC provides for only two levels of geometry: cell compartments and collections of these compartments into entire cells. Each cell compartment maintains its own probabilistic transition tree, which, although may be somewhat redundant, is what allows SPC to exhibit localized concentrations of proteins similar to real biological systems. It is the artificial gradients computed between adjacent cell compartments (and between the aggregate cells) that drive the diffusion of species across compartment boundaries.

A side effect of the lossy encoding of protein data within SPC is that the actual protein-protein interactions must be simulated in a special way: Each time step, the cell compartment queries its local transition probability tree and generates a list of plausible proteins (that is, proteins that the data structure suggests are likely, all constructed from valid $n$-grams). This sampling window of proteins is evaluated; instruction proteins are interpreted, and if they can bind to other proteins in the window, then they are allowed to do so. The result of evaluating these proteins can be one of these outcomes:

- A new protein might be introduced, such as the earlier example in which two smaller proteins are fused into a new product.
- One of the existing proteins might have been destroyed; again, the earlier example showed this because the "l" and "a" proteins were consumed to form the new fusion protein.

When either of these events occurs, the local data structure is updated to reflect the change in density in $n$-grams. The net result is that the local, lossy store of $n$-grams is sampled, their interactions are computed, and the $n$-grams store is updated. This allows the distribution of plausible proteins to evolve over time. Once the diffusion between cell compartments is added, this provides for the system-wide mobility of protein species and sharing of intermediate work products.

Another substantial impact of this design is nondeterminism, meaning that two subsequent runs of the simulation will probably produce (hopefully statistically insignificantly) different results. At each time step, there is stochasticity (pseudorandomness) in which plausible proteins will be generated by the local data store; how they interact will guide the evolution of these probabilistic transition trees, which amplifies the results of these relatively minor changes across time. Ultimately, these random fluctuations will be observable in the final assay of proteins across the entire simulation. For large, complex problems that are searching for acceptable choices in huge search spaces, this element of nondeterminism may be acceptable; when computing how much liquid oxygen should be fed to the space shuttle engines this microsecond, nondeterminism is undesirable.

One of the key limiting factors in this kind of simulation is the model of the underlying chemistry that must be used. Within SPC, this model is necessarily very crude. It is not based on MD so much as computing dynamics, reflecting the fact that—as of now—it is a much more realistic simulation of a new computing environment than it is a simulation of how proteins interact.

### 12.4.2.3  Computing Power

The SPC architecture implicitly supports a peer-to-peer model of computing, rather than a top-down approach, because the cells and their compartments are only directly affected by their immediate neighbors. This means that each cell compartment can be modeled on one core of a multicore processor or could, in fact, be modeled as one cell compartment per computer on a network; the mechanics of mapping cell compartments to silicon processors is independent of the computing work being performed.

More importantly, it should be recognized that all programs written within the SPC framework are automatically separable. Unlike traditional von Neumann computing, in which code fragments must be manually separated and farmed out to different processors, each SPC cell compartment is entirely independent and has autonomy in performing its own calculations. The localization of the results is what allows for the exploration of many different portions of the solution space simultaneously; the diffusion of intermediate products across compartment and cell boundaries is what allows for cross-location correction and refinement of results. The distribution of work will not be as cleanly partitioned within SPC as it will be in methods where the task is manually decomposed into nonoverlapping subproblems, but the ability to scale the environment up to as many processors as happen to be available *without* any additional coding required is, in fact, significantly different.

### 12.4.2.4  Future Goals

For as long as it remains a simulation within a digital computer, SPC will still be subject to many of the same shortcomings. It is the long-term hope of the project that there will be a convergence between the biological and computational sciences and that as the protein-interaction databases and our understanding of protein folding improve, so will our ability to model proteins and chemistry more faithfully within the computer. In this way, the biological sciences increase in capability until it is possible to perform massive computational searches in vivo. At the same time, the SPC simulations ought to have improved so that it becomes merely a kind of design/ testing platform that allows you to debug your protein programs before going to the effort of introducing them into the reaction environment.

Ultimately, the advantage in using this method is the long-term result of being able to encode real-life problems in biological proteins, introduce them into suitable receptor cells, and after some time has elapsed—representing CPU-years and more memory than is currently available—assay the results to read the answer out of the new distribution of protein products.

## 12.5  SUMMARY

Simulation is an art that requires a careful eye for both detail and the higher-level abstractions of what properties and behaviors best define a system. It is possible to have two very different simulations claim to represent the same root process simply because they focus on different aspects of that common process. In this chapter, we have reviewed the nature and motivation behind simulation, explored some of the issues that make good simulation challenging, and stepped through two very different simulation projects: Rosetta@Home and SPC. These two products highlight both the simulation of biological systems as well as the use of simulations inspired by biology. As processing power increases, researchers will become increasingly dependent upon simulations of all kinds to perform the work that would otherwise be too expensive or time-consuming.

## 12.6  PROBLEMS

1. Name three other projects that rely on simulation to help accomplish core research. What do these projects have in common? What are the central processes on which each is focused, and what details have they both included and elided to accomplish their purposes?
2. Is Conway's Game of Life a simulation or a mere mathematical curiosity? Why? Under what circumstances or assumptions? What about genetic algorithms?
3. Complex adaptive systems, such as Axelrod's study of the iterated prisoner's dilemma often take the form of simulations. What would be the advantages and pitfalls of using simulation in this context?
4. What is ergodicity, and how might it influence simulation design, implementation, and analysis? Address the same issues with respect to Lyapunov functions.
5. In what way would a sensitive dependence on initial conditions be expected to serve or confound simulations of protein folding?
6. Is SPC ever likely to mature sufficiently that it can be either a useful platform for computing or a useful simulation of real protein interactions? Why?
7. What is the average CPU load on your machine at work? At home? What opportunities exist for these cycles to be harvested? What are the best applications to take advantage of this resource?

## REFERENCES AND NOTES

Anderson, David P., and Gilles Fedak, "The Computational and Storage Potential of Volunteer Computing." *IEEE/ACM International Symposium on Cluster Computing and the Grid*, Singapore, May 16–19, 2006. Available at http://boinc.berkeley.edu/boinc_papers/internet/paper.pdf.

Anderson, T. E., D. E. Culler, and D. Patterson, "A Case for NOW (Networks of Workstations)." *IEEE Micro*, 15 (February 1995): 54–64.

Baker, D., "Rosetta@Home: Protein Folding, Design and Docking." Available at http://boinc.bakerlab.org/rosetta/.

BOINC, "Open-Source Software for Volunteer Computing and Grid Computing." Available at http://boinc.berkeley.edu/.

Cui, Y., R. S. Chen, and W. H. Wong, "Protein Folding Simulation with Genetic Algorithm and Supersecondary Structure Constraints." *Proteins,* 31 (May 15, 1998): 247–257.

Eichelberger, C., and K. Najarian, "Simulating Protein Computing: Character Recognition via Probabilistic Transition Trees." *Proceedings of IEEE Granular Computing* (2006): 101–105.

Goldstine, H. H., and J. von Neumann, "On the Principles of Large Scale Computing Machines." In *John von Neumann Collected Works*, ed. A. H. Taub, vol. V, 1–32. New York: Macmillan.

Hoare, Tony, and Robin Milner, eds., *Grand Challenges in Computing Research*. The British Computing Society. Available at http://www.ukcrc.org.uk/gcresearch.pdf.

IBM Blue Gene Team: F. Allen, G. Almasi, W. Andreoni, D. Beece, B. J. Berne, A. Bright, J. Brunheroto, C. Cascaval, J. Castanos, P. Coteus, P. Crumley, A. Curioni, M. Denneau, W. Donath, M. Eleftheriou, B. Fitch, B. Fleischer, C. J. Georgiou, R. Germain, M. Giampapa, D. Gresh, M. Gupta, R. Haring, H. Ho, P. Hochschild, S. Hummel, T. Jonas, D. Lieber, G. Martyna, K. Maturu, J. Moreira, D. Newns, M. Newton, R. Philhower, T. Picunko, J. Pitera, M. Pitman, R. Rand, A. Royyuru, V. Salapura, A. Sanomiya, R. Shah, Y. Sham, S. Singh, M. Snir, F. Suits, R. Swetz, W. C. Swope, N. Vishnumurthy, T. J. C. Ward, H. Warren, and R. Zhou, "Blue Gene: A Vision for Protein Science Using a Petaflop Supercomputer." *IBM Systems Journal: Deep Computing for the Life Sciences,* 40 (2001): 2. Available at http://researchweb.watson.ibm.com/journal/sj/402/allen.html.

*IBM Journal of Research and Development*, 49 (2005): 2/3. Available at http://www.research. ibm.com/journal/rd49-23.html.

Pande, V., *Folding@Home*. Available at http://folding.stanford.edu/English/Main.

Pedersen, J. T., and J. Moult, "Protein Folding Simulations with Genetic Algorithms and a Detailed Molecular Description." *Journal of Molecular Biology,* 269 (June 6, 1997): 240–259.

Simons, K. T., C. Kooperberg, E. Huang, and D. Baker, "Assembly of Protein Tertiary Structures from Fragments with Similar Local Sequences Using Simulated Annealing and Bayesian Scoring Functions." *Journal of Molecular Biology*, 268 (April 25, 1997): 209–225.

Simons K. T., I. Ruczinski, C. Kooperberg, B. A. Fox, C. Bystroff, and D. Baker, "Improved Recognition of Native-like Protein Structures Using a Combination of Sequence-Dependent and Sequence-Independent Features of Proteins." *Proteins*, 34 (January 1, 1999): 82–95.

*The Rosetta Algorithm*. CCR Yeast Resource Center, University of Washington. Available at http://depts.washington.edu/yeastrc/pages/psp_rosetta.html.

*The Rosetta Commons*. Available at http://www.rosettacommons.org/.

# 13 Software, Databases, and Other Resources for Systems Biology

## 13.1 INTRODUCTION AND OVERVIEW

In the last few decades, the molecular biology and the equipment available for research in this field have been advanced significantly. Therefore, large sequencing projects, in which the entire genetic sequence of an organism is obtained, are now routine. Today, several bacterial genomes, as well as those of some simple eukaryotes (e.g., *Saccharomyces cerevisiae*, or baker's yeast) and more complex eukaryotes (*Caenorhabditis elegans* and *Drosophila*) have been sequenced. The Human Genome Project is another example of rapidly growing research that provides a huge amount of data. These large amounts of sequence data can be used for

- Analysis of the organization of genes and genomes and their evolutionary processes
- Prediction of the function of newly identified genes
- Protein sequence prediction from DNA sequence
- Identifying regulatory factors of a specific gene or RNA
- Identification of mutations that cause diseases and others

Inasmuch as this information can help scientists discover or predict some unknown features, bioinformatics provides some methods for recording, annotating, searching/retrieving, analyzing, and storing nucleic acid sequence (genes and RNAs), protein sequences, and structural information.

Computational biology is a new branch that is related to these processes. It involves the following subjects:

- Finding the DNA gene sequences of various organisms
- Developing methods to predict the structure and/or function of newly discovered proteins and sequences
- Clustering protein sequences into families of related sequences and predicting protein models
- Aligning similar proteins and developing phylogenetic trees to examine evolutionary relationships

In this chapter, we will discuss fundamental subjects that are needed to understand these new fields.

## 13.2   DATABASES

Storing the information is the first step in managing and analyzing it. Hence, bioinformatics provides biological data stores and retrieval systems to support data and
help researchers to use these data. A simple data store might be a file that contains
many sequences and annotations. For example, a record that contains nucleotide
sequence information includes the input sequence that describes the type of molecules, the particular organism's scientific name, and often the articles that have used
these sequences.

There are different types of sequences with different types of information. Some
of these include nucleotide databases, protein sequence databases, macromolecular
structure databases, array expression databases, and databases of expressed sequence
tags. In this chapter, we will deal with the two first databases.

### 13.2.1   Nucleotide Databases

Nucleotide sequences are supported by different databases, but recently, there has been
consolidation of the data into four International Nucleotide Sequence Databases.

#### 13.2.1.1   GenBank

GenBank is the NIH genetic sequence database, a collection of all publicly available
DNA sequences with annotations. This database is updated every 2 months. Each
GenBank entry includes a brief description of the sequence, the scientific name,
and taxonomy of the particular organism from which the data is isolated and a table
of features that identifies coding or noncoding regions and other sites of biological significance, such as transcription units, sites of mutations or modifications, and
repeats. Protein translations for coding regions are included in the feature tables.
Bibliographic references are included for all known sequences with a link to the
Medline unique identifier: http://www.ncbi.nlm.nih.gov/Genbank/.

#### 13.2.1.2   DNA Data Bank of Japan

The DNA Data Bank of Japan (DDBJ) began DNA data bank activities in 1986.
Because an organism's evolution can be seen more directly with DNA sequence
records than other biological materials, the DNA sequence records are a valuable
measurement for the researchers. The DDBJ provides online access for all researchers: http://www.ddbj.nig.ac.jp/.

#### 13.2.1.3   European Molecular Biology Laboratory

The European Molecular Biology Laboratory (EMBL) Nucleotide Sequence
Database (also known as EMBL-Bank) is a resource for primary nucleotide
sequence. The main sources for DNA and RNA sequences are accessible through
http://www.ebi.ac.uk/embl/.

#### 13.2.1.4   International Sequence Database

All introduced databases are gathered together in the international sequence database.
The DDBJ, the EMBL, and GenBank at the National Center for Biotechnology put their
information in this database. These three organizations exchange data on a daily basis.

This collaboration that exists in the International Nucleotide Sequence Databases has influenced many research projects. The database (http://www.insdc.org/) presents the goals and policies of this useful collaboration and also has links to the three partners' databases and retrieval tools.

## 13.2.2  PROTEIN DATABASE

A protein can be analyzed in the laboratory, examining its sequence or structure. In this section, we study the sequence databases.

The amino acid sequence—primary structure—is unique to each protein. This defining sequence can be obtained from a DNA sequence, which is translated into the amino acid sequence of the particular protein of our interest.

Sometimes, only portions of a particular protein are sequenced and can be used to identify the expressed protein. The expressed proteins are additionally annotated in a database such as Swiss-Prot and the Protein Information Resource (PIR). These databases have collaborated to form the Universal Protein Resource (UniProt) database.

### 13.2.2.1  Swiss-Prot

Swiss-Prot, jointly maintained by the Swiss Institute for Bioinformatics and the European Bioinformatics Institute, describes itself as "a curated protein sequence database which strives to provide a high level of annotation (such as the description of the function of a protein, its domains structure, post-translational modifications, variants, etc.), a minimal level of redundancy and high level of integration with other databases."

### 13.2.2.2  Protein Information Resource

The PIR is located at Georgetown University Medical Center. PIR was established in 1984 by the National Biomedical Research Foundation and has been used as a resource that can help researchers in the identification and interpretation of protein sequence information (http://pir.georgetown.edu).

### 13.2.2.3  Translated EMBL

Translated EMBL has the translations of all coding sequences present in the DDBJ/EMBL/GenBank, Nucleotide Sequence Database, and also protein sequences extracted from the literature or submitted to UniProtKB/Swiss-Prot. This is not yet integrated into UniProtKB/Swiss-Prot (http://www.ebi.ac.uk/trembl/index.html).

### 13.2.2.4  Universal Protein Resource

UniProt is the world's most comprehensive repository of protein sequence and function created by joining the UniProtKB/Swiss-Prot, UniProtKB/TrEMBL, and PIR information.

UniProt consists of three components, each specialized for a specific use. The UniProt Knowledgebase (UniProt) is the central access point for extensive protein information, including function, classification, and cross-reference. The UniProt Reference Clusters (UniRef) combines related sequences closely into a single record in order to have fast search ability. The UniProt Archive (UniParc) is a comprehensive repository, which gathers the history of all protein sequences (http://www.expasy.uniprot.org/).

## 13.3 SCORING MATRIX

The basic idea of the sequence alignment programs is to align two sequences to produce the highest score that displays the highest similarity between sequences. A scoring matrix is a similarity matrix that is used to add points to the score for each similarity and subtract for each mismatch.

An important and meaningful tool for evaluating the results of a pairwise sequence alignment is the "substitution matrix," which assigns a match score for aligning any possible pair of residues in respect to their history of alignment. In general, different substitution matrices are introduced for finding similarities between sequences.

In bioinformatics, scoring matrices for computing alignment scores are often related to observed substitution rates. The nucleic acid scoring matrices commonly involve simple match/mismatch scoring schemes, while the protein alignment matrices are more complex, with scores designed to display similarity between the different amino acids rather than simply scoring identities given from the substitution frequencies seen in multiple alignments of sequences. In fact, for every substitution occurring in real sequences, a substitution score, which reflects the rate of substitution between two particular amino acids, is assigned. In other words, these substitution values reflect estimates of the naturally occurring probability of these two amino acids being interchanged. The other issue that is noticeable is the frequency of occurrence of a special amino acid. The higher scores reflect the rate of relatedness of two special amino acids and also predict that the replacement between those amino acids is not taken by chance and is an evolutionary phenomenon, but the low substitution scores reflect that the two particular amino acids are replaced by chance. Thus, equal sequences without substitutions of amino acids are assigned the most positive scores. Frequently observed substitutions also receive positive scores, but matches that are not evolutionary substitutions and so have not been seen frequently in nature are given negative scores.

The two most commonly used types of protein scoring matrices are the *point accepted mutation* (PAM) and *blocks substitution matrix* (BLOSUM).

### 13.3.1 Point Accepted Mutation

PAM approach to scoring was first introduced by Dayhoff and coworkers in the 1970s. PAM matrices are based on global alignments of closely related proteins.

The PAM is the matrix achieved from comparison of similar sequences. The similar sequences do not have any mismatches or have a maximum of 1% mismatch regions in the whole sequences. A PAM matrix is specific for a particular evolutionary distance but may be used to generate matrices for greater evolutionary distances by multiplying it repeatedly by itself. However, at large evolutionary distances, the information that is presented in the matrix is essentially degenerate. It is rare that a PAM matrix would be used for an evolutionary distance greater than 256 PAMs.

### 13.3.2 Blocks Substitution Matrix

The BLOSUM compares the blocks of sequences that are derived from the database blocks. BLOSUM is based on local multiple alignments of more distantly related sequences.

The BLOSUM62 is a default blocks database matrix. This matrix is calculated from alignment of sequences that are more than 62% identical. This nomenclature is also useful for calling other BLOSUM matrices. For example, the BLOSUM 80 matrix would be derived from a set of sequences having 80% sequence identity. Unlike PAM matrices, new BLOSUM matrices cannot be derived from old BLOSUM matrices.

For a set of sequences, the level of relatedness is useful for choosing the most appropriate scoring matrix for aligning the set. Comparison of closely related sequences typically would use BLOSUM matrices with higher numbers and PAM matrices with lower numbers because they have punished the mismatch words more than normal matrices, which are useful in dealing with less similar sequences. Conversely, BLOSUM matrices with low numbers and high-value PAM matrices are preferred for comparison of distantly related proteins. So, one cannot use a single matrix for aligning a huge range of different sequences.

## 13.4  ANALYSIS SOFTWARE

Because of the growing rate of finding new sequences in both nucleotide and amino acids, there is an urgent need for appropriate software that can analyze these new data and align them with known databases. The dynamic programming method of Needleman and Wunsch would not be appropriate because it checks all possible alignments and is a time-consuming method. There are many hierarchical methods for aligning and searching these new sequences. Two powerful methods that are used frequently and available online are FASTA and BLAST.

### 13.4.1  FASTA

Pearson and Lipman (1988) developed an algorithm called FASTA, which provides a rapid way to find a good, if short, similar subsequence between a new sequence and any sequence in a database. Now, it can also be used for sequence similarity searching against complete proteome or genome databases.

#### 13.4.1.1  FASTA Algorithm

FASTA is a program for rapid alignment of pairs of protein and DNA sequences. The FASTA algorithm contains four steps.

Step 1:  The first step identifies regions shared by the two sequences with the highest density of identities, rather than comparing individual words in two sequences. In other words, FASTA searches for *k-tup*, which is the matching sequence pattern. The k-tup parameter determines how many repeated identities are required in a match. A k-tup value of 2 is frequently used for protein sequence comparison, which means that the program examines only those portions of the two sequences that have at least two identical residues in both sequences. More sensitive searches can be done using k-tup = 1. For DNA sequence comparisons, the k-tup parameter can range from 1 to 6.

The *diagonal* method is used to find all regions of similarity between the two sequences, applying k-tup matches and penalizing for the occurrences of mismatches. This method identified regions of a diagonal that have the highest density of k-tup matches. The term *diagonal* refers to the diagonal line that is seen on a dot matrix plot when a sequence is compared with itself, and it denotes an alignment between two sequences without gaps.

FASTA saves the 10 best local regions, whether they are on the same or different diagonals.

Step 2:  After the 10 best local regions are found in the first step, they are rescored using a scoring matrix that allows runs of identities shorter than k-tup residues and conservative replacements to contribute to the similarity score. For protein sequences, this score is usually calculated using the PAM250 matrix.

FASTA can also be used for DNA sequence comparisons, and matrices can be constructed that allow separate penalties for transitions. For each of the best diagonal regions rescanned with the scoring matrix, a subregion with the maximal score is identified. Initial scores are used to rank the library sequences. These scores are referred to as init1 score.

Step 3:  FASTA checks, during a library search, to see whether several initial regions can be joined together in a single alignment to increase the initial score. FASTA calculates an optimal alignment of initial regions as a combination of compatible regions with maximal score. This optimal alignment of initial regions can be rapidly calculated using a dynamic programming algorithm.

FASTA uses the resulting score, referred to as the initn score, to rank the library sequences. The third "joining" step in the computation of the initial score increases the sensitivity of the search method because it allows for insertions and deletions as well as conservative replacements. The modification does, however, decrease selectivity. The degradation selectivity is limited by including in the optimization step only those initial regions whose scores are above an empirically determined threshold: FASTA joins an initial region only if its similarity score is greater than the cutoff value, a value that is approximately 1 standard deviation above the average score expected from unrelated sequences in the library. For a 200-residue query sequence and k-tup-2, this value is 28.

Step 4:  FASTA constructs Needleman-Wunch-Sellers algorithm optimal alignment of the query sequence and the library sequence, considering only those residues that lie in a band 32 residues wide centered on the best initial region found in Step 2. FASTA reports this score as the optimized (opt) score. After a complete search of the library, FASTA plots the initial scores of each library sequence in a histogram, calculates the mean similarity score for the query sequence against each sequence in the library, and determines the standard deviation of the distribution of initial scores. The initial scores are used to rank the library sequences, and in the fourth and final step of the comparison, the highest scoring library sequences are aligned using a modification of the standard Needleman-Wunch-Sellers optimization method. The optimization employs the same scoring matrix used in determining the initial regions; the resulting optimized alignments are calculated for further analysis of potential relationships, and the optimized similarity score is reported.

### 13.4.1.2   FASTA Variants

Different versions of FASTA are produced to align sequences. They are described below.

FASTA: This program compares a protein sequence to another protein sequence or to a protein database and a DNA sequence to another DNA sequence or a DNA library.

SSEARCH: This program aligns a protein sequence to another protein sequence or a DNA sequence to another DNA sequence using rigorous Smith-Waterman alignment. It can be used for searching among a protein database or a DNA library. The speed of this program is very slow.

FASTX/FASTY: This program compares a DNA sequence to a protein sequence database, translating the DNA sequence in three forward (or reverse) frames and allowing frame shifts.

TFASTX/TFASTY: This program compares a protein sequence to a DNA sequence or DNA sequence library. It translates the DNA sequence in three forward and three reverse frames, and the protein query sequence (a sequence that is searched among the database) is compared with each of the six derived protein sequences.

FASTS/TFASTS: This program is used to compare a set of short peptide subsequences, as would be obtained from mass spectrometry. FASTS compare protein sequences, and TFASTS searches the DNA database.

LALIGN/PLALIGN: This program can align two sequences of proteins for finding the locally similar regions between them. LALIGN aligns several sequence alignments if there are several similar regions. LALIGN can identify similarities due to internal repeats or similar regions that cannot be aligned by FASTA because of gaps. PLALIGN, which looks much like a "dot plot," visualizes the alignment by plotting a graph of the sequence alignments.

PRSS/PRFX: This program uses a Monte Carlo analysis for evaluating the significance of pairwise similarity scores. Similarity scores for the two sequences are calculated, and then the second sequence is rearranged randomly 200–1,000 times, and the alignment task with the first sequence is repeated. PRSS can use one of two shuffling strategies. One strategy simply keeps the amino acid composition of the rearranged sequence identical to the primary one. The second destroys the order but keeps the composition of small (10–25 residues) subsequences of the rearranged sequence. PRFX does a similar rearranging method but compares a translated DNA sequence to a protein sequence using the FASTX algorithm.

GARNIER/CHOFAS: This program is designed to predict protein secondary structure using its sequence.

## 13.4.2 BLAST

BLAST was developed by Altschul et al. in 1990. It is an abbreviation for Basic Local Alignment Search Tool. This method is used to compare a query sequence with those gathered in nucleotide and protein databases by aligning the query sequence with previously characterized genes. The emphasis of this tool is to find regions of sequence similarity, which will yield functional and evolutionary signs about the structure and function of this query sequence. Regions of similarity detected via this type of alignment tool can be either global, where regions of similarity can be detected across otherwise unrelated genetic code, or local, where the region of similarity is based in subsequences of query sequence.

### 13.4.2.1 BLAST Algorithm

The BLAST algorithm is a heuristic search method that seeks subsequences of length $W$ (default = 3 in BLASTP) that score at least $T$ when aligned with the query and scored with a substitution matrix. Words in the database that score $T$ or greater are extended in both directions in an attempt to find a locally optimal ungapped alignment or high-scoring pair (HSP) with a score of at least $S$ or an $E$ value lower than the specified threshold. HSPs that meet these criteria will be reported by BLAST, provided they do not exceed the cutoff value specified for number of descriptions and/or alignments to report. This algorithm has three steps:

1. In the first step, it aligns query subsequences to find a list of high-scoring words. A list of words of length three in the query protein sequence is made starting with positions 1, 2, and 3; then 2, 3, and 4; etc., until the last three available positions in the sequence are reached (word length 11 for DNA sequences, 3 for programs that translate DNA sequences). Then, using the BLOSUM62 substitution scores, the query sequence words in Step 1 are aligned for finding an exact match with all of the database words of that particular length.

   The remaining high-scoring words that gathered possible matches to each three-letter position in the query sequence are organized and used to generate a significant search tree that compares the sequences rapidly to those in the database.
2. Then, the 50 ordered words of query are aligned with the sequence. If a match is found, this match is used to seed a possible ungapped alignment between the query and database sequences.
3. In the original BLAST method, an attempt was made to extend an alignment from the matching words in each direction along the sequences, continuing for as long as the score continued to increase, as illustrated below. The extension process in each direction was stopped when the resulting summed score reached its maximum. At this point, a larger stretch of sequence (called the HSP or high-scoring segment pair), which has a larger score than the original word, may have been found.

L P P Q G L L QUERY SEQUENCE
M P P E G L L DATABASE SEQUENCE
<WORD> THREE LETTER WORD FOUND
INITIALLY
7 2 6 BLOSUM62 scores, word
< HSP > HSP SCORE = 9 + 15 + 8 = 32

In the later version of BLAST, called BLAST2 or gapped BLAST, a different and much more time-efficient method is used. The method starts by making a list of high-scoring matching words, as in Steps 1–4 above, with the exception that a lower value of $T$, the word cutoff score, is used.

In the next state, significance of HSP is calculated by BLAST. A probability that two random sequences, having the length of query and database sequences (which is approximately equal to adding the lengths of all of the database sequences together), could achieve the HSP score is calculated. The probability $p$ of seeing a score $S$ equal to or greater than $x$ is given by the equation

$$p(S > x) = 1 - \exp\left(-e^{-\text{EMBED Equation.3}(-x_u)}\right) \tag{13.1}$$

where $u = [\log (Km'n')]/\text{EMBED Equation.3}$ and $K$ and EMBED Equation.3 are parameters that are calculated by BLAST for the amino acid substitution scoring matrix, $n'$ is the effective length of the query sequence, and $m'$ is the effective length of the database sequence.

The effective length of a sequence is a number that reflects the nonsimilar regions of the sequence and calculated by subtracting the average length of an alignment between two random sequences of the same length from the actual lengths of the query and database sequences. $m'$ and $n'$ are calculated from the following relationship:

$$m' = m - (\ln Kmn / H)$$
$$n' = n - (\ln Kmn / H) \tag{13.2}$$

where $H$ is the average expected score per aligned pair of residues in an alignment of two random sequences. $H$ is calculated from the relationship

$$H = (\ln Kmn) / l \tag{13.3}$$

where $l$ is the average alignment length of sequences with the actual lengths of $m$ and $n$, using the same scoring system as used in the database search; $l$ is measured from actual alignments of random sequences. The basis in using these reduced lengths in statistical calculations is to permit a local alignment occurrence in which an alignment starting near the end of one of the sequences can reach to an optimal alignment.

Smith-Waterman local alignments can be used for the query sequence with each of matched sequences in the database. Earlier versions of BLAST produced only ungapped alignments that included the initially found HSP. If two HSPs were found, because of impossible alignment between two regions without gaps, two separate alignments were produced. BLAST2 version performs the search with an additional issue. It performs a single alignment with gaps that can include the entire primary-found HSP regions. Aligning of sequences may be divided into subalignments of the sequences, one starting at some point in sequence 1 and going to the beginning of the sequences and another starting at the distal ends of the sequences and ending at the same position in sequence 1. A similar method is used to produce an alignment starting with the alignment between the central pair in the highest-scoring region of the HSP pattern as a seed for producing a gapped alignment of the sequences. After that state, the alignment score is calculated, and the significance of alignment for those sequences is achieved. When the score for a given database sequence satisfies the user-selectable threshold parameter $E$, the match is reported.

### 13.4.2.2 BLAST Variants

PHI-BLAST (Pattern-Hit-Initiated BLAST): This is a search program that uses local alignments surrounding the match for matching of regular expressions.

PSI-BLAST: This program is a repeatedly searching method to find more distant matches to a test protein sequence. It can be done for additional sequences that match an alignment of the query and initially matched sequences.

blastp: This program compares an amino acid query sequence with a protein sequence database.

blastn: This program compares a nucleotide query sequence against a nucleotide sequence database.

blastx: This program compares a nucleotide query sequence against a protein sequence database. One could use this option to find potential translation products of an unknown nucleotide sequence.

tblastx: Compares the six-frame translations of a nucleotide query sequence against the six-frame translations of a nucleotide sequence database. Note that the tblastx program cannot be used with the nr database on the BLAST Web page.

## 13.5   BIOINFORMATICS IN MATLAB

The Bioinformatics Toolbox offers computational molecular biologists and other researchers a wide environment to find ideas, use new algorithms, and build applications in drug research, genetic engineering, and other genomics and proteomics projects. The toolbox provides access to genomic and proteomic data formats, analysis techniques, and specialized visualizations for genomic and proteomic sequences. Most functions are used in the open MATLAB® language.

### 13.5.1   SEQUENCE ANALYSIS

The Bioinformatics Toolbox provides functions for genomic and proteomic sequence analysis and visualization of alignment between sequences. Analyses can change from multiple sequence alignments to building and interactively viewing and manipulating phylogenetic trees.

### 13.5.2 SEQUENCE ALIGNMENT

The Bioinformatics Toolbox offers a comprehensive set of analysis methods for performing pairwise sequence, sequence profile, and multiple sequence alignment. MATLAB implements standard algorithms for local and global sequence alignment, such as the Needleman-Wunsch, Smith-Waterman, and profile-hidden Markov model algorithms.

## 13.6 SUMMARY

This chapter introduces the resources available to bioinformaticians and systems biologists. These resources include software tools and databases that provide invaluable help for both research work and clinical decision making.

## 13.7 PROBLEMS

1. Find out what is known about human CYP2B genes by searching the NCBI databases.
2. Compare the frog Rhodesian cDNA against the mouse genomic DNA. Use different word sizes.
   a. What do you notice?
   b. Which word sizes give the clearest plot?
3. Make local and global sequence alignment for the coding sequence of Y12618 and ATU89272. Compare the results.
4. Search OsHT01 nucleotide sequence (http://www.ncbi.nlm.nih.gov/entrez/viewer.fcgi?val=AJ557777.1 AJ557777) against the NCBI NR database by BLAST.
5. Search cytokine-induced protein sequence (http://www.ncbi.nlm.nih.gov/entrez/viewer.fcgi?db=protein&val=32129199 NP_149073) against the NCBI NR database by FASTA.

## REFERENCES

Altechul, S. F., W. Gish, W. Miller, W. Myers, and D. J. Lipman, "Basic local alignment search tool," *J. Mol. Biol.*, 215(1990): 3, 403–410.

Pearson, W. R., and D. J. Lipman, "Improved tools for biological sequence comparison" *Proc. Natl. Acad. Sci. U.S.A.*, 85(1998): 8, 2444–2448.

# 14 Future Directions

## 14.1 INTRODUCTION AND OVERVIEW

Having covered the basic methods and approaches to systems biology in the preceding chapter, it is appropriate to focus on the areas of research that are currently active to extend our capabilities. The presentation begins on a small scale—how to perform single-cell microarray assays—then scales up to high-throughput methods of protein assay, jumps to ways to integrate low-level data with higher-level abstractions of biological process, addresses the core question of how to make the association between low-level genetic data and high-level phenotypes, and concludes with a review of molecular-level imaging and the systems biology approach.

## 14.2 SINGLE-CELL MICROARRAY AND SYSTEMS BIOLOGY

The advantages of a holistic, systems approach to biology are offset by the complications: There are so many active entities whose interactions are so varied and indirect that disentangling the various influences—identifying, quantifying, and delimiting them—can be intractable. It would be a difficult problem even if the data fully describing all of the myriad factors were available, but they are not, so one of the first issues a research needs to address is "What data do I need, and what data can I get, to help solve my problem?" Typically, this is an iterative process, requiring many rounds of review, collection, and analysis. The heart of investigations is how to isolate the key processes under consideration from the large amount of noise present. In this case, "noise" refers to the volumes of background data that correspond to other biological processes tangential to the central process. For example, if you wanted to study the gene activation network involved during glycolysis, then the activation level of genes controlling cell division are probably not relevant and represent data that ought to be discarded. Single-cell microarrays are a way to isolate key data from a larger biological context, either by focusing on cell selection (a hardware and process approach) or on the amplification of genetic data within a single cell (a purely process-oriented approach).

### 14.2.1 MICROARRAYS TO IDENTIFY SINGLE CELLS

The first of these methods, identifying one specific type of cell in a larger collection of cells as used by (Tajiri et al., 1997), is illustrated in Figure 14.1. The task they undertook was to identify which lymphocytes were receptive to a specific antigen. To do this, a group of live cells are deposited onto a special microarray that contains multiple chambers, each of which is designed to contain a single cell (A); to eliminate false-positives, a solution of nonspecific, tagged proteins is introduced, and those cells that bind to these nonspecifics are eliminated from consideration (B);

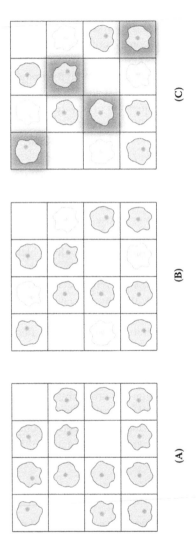

**FIGURE 14.1**    Three phases of using a single-cell microarray to isolate one specific cell type.

lastly, the fluorescence-labeled antigens are applied to the matrix of cell chambers, and those cells that bind to the antigens are identifiable (C). These researchers also employed a tool that can extract cells from individual wells using microscopy, allowing them to harvest exclusively the positive cells.

The benefits of this approach include

1. The proportion of cells receptive to a certain membrane-binding factor can be quantified within a larger sample of cells. This supports experiments into the functional operation of cells as black boxes.
2. When, as was the case for these researchers, the proportion of cells likely to be positive cases (receptive to the specific antibody) is known to be low, it represents a way to filter out the large proportion of noise (noninteresting cells).

In this approach, the key differentiating factor is the hardware, both the microarray that contains the wells into which the cells are deposited and isolated from their peers and the manipulator that allows the positively identified cells to be extracted. In essence, this is a quantification and a purification technique, albeit one that produces an extremely small—if valuable—harvest. Continued improvement in the hardware technology should allow both for higher volume analysis as well as greater automation and discriminatory power in the method.

### 14.2.2 MICROARRAY ANALYSIS OF A SINGLE CELL

The second use of microarrays to investigate single cells is purely a process-oriented approach that concerns how best to amplify the contents of a single cell to a level that makes it a candidate for analysis on a traditional microarray. Esumi et al. (2008) studied two techniques of addressing this issue: T7 RNA transcription (T7) and reverse transcription–polymerase chain reaction (RT-PCR).

T7 is an RNA transcription method documented by Sastry and Ross (1997). The polymerase is derived from bacteriophages (viruses, of which HIV is a representative member that employs T7-type transcription) and uses T7 promoter regions to regulate transcription. The key advantage of T7, according to Esumi et al. (2008) is the ability to replicate RNA fragments with relatively little bias. That is, the proportion of different species of RNA strands appears to be preserved even over multiple rounds of replication. Unfortunately, there is a commensurate cost in a reduced length of the residues copied.

RT-PCR is also a method rooted in the behavior of phages, since the reverse transcription is the method that a virus employs to have its own RNA produce DNA within the host cell. RT-PCR enjoys the benefit of an exponential increase in the amount of copied residue (much like T7). The Clontech Super SMART system used by Esumi et al. (2008) is based on the Moloney murine leukemia virus reverse transcriptase that allows relatively long residues to be copied reliably. Unfortunately, RT-PCR appears to alter the distribution of residues in the amplified sample (as per Iscove et al., 2002).

The key to Esumi et al. (2008) and other research is the recognition that these methods are not mutually exclusive. The challenge then becomes balancing the strengths and weaknesses of these two amplification techniques so that the resulting residue products are as representative of the contents of the starting cell as possible. This becomes easier with detailed understanding of the mechanics of the specific cell under study and is similarly more difficult when trying to achieve a general-purpose means of allowing traditional microarray technology to be applied to the amplified contents of a single cell.

## 14.3   HIGH-THROUGHPUT PROTEIN ASSAYS AND SYSTEMS BIOLOGY

Voice my opinion with volume

**Public Enemy**
*Rebel Without a Pause*

Gel electrophoresis—the ubiquitous sodium dodecyl sulphate–polyacrylamide gel electrophoresis (SDS-PAGE)—is a way to separate proteins so that they can be analyzed, but it is time-consuming and limited to only a relatively small number of proteins per run and is not particularly useful when the constituent proteins are unknown because it provides what are essentially only benchmarking data on the relative size of the residues in the wells. Western blots are more useful for testing the presence of specific proteins (and typically involve an early SDS-PAGE step) but are again limited by the number of proteins for which antibodies are known (and prepared) and are typically for confirmation rather than exploration. Given the myriad proteins present within a single cell, a systems biology approach requires a much broader, reproducible, higher-volume method of identifying and quantifying proteins. Traditional methods share these limitations:

Because they were novel, they were initially conceived as single-sample or low-volume methods and generally take a few hours to complete a single run.

They are manually intensive and hence susceptible to operator influence. This introduces a level of variability in the individual results.

High-throughput protein assays are generally techniques that rely on increasing the number of tests that can be run in parallel through a single system or machine, typically in a highly automated manner. Assay tasks that can be scaled up in this manner include

- High-throughput screens are significantly automated and use plates that are multiples of 96-wells to conduct up to tens of thousands of assays per day, allowing for a broad-spectrum approach to protein quantification.
- ELISA that can detect the presence of antigens or antibodies, typically administered on 96-well plates.
- SDS-PAGE–type protein separation in which the presence (or absence) of known proteins is to be confirmed (Caliper LifeScience's LabChip 90).
- Enzymatic assays in which different candidate inhibitors are to be tested simultaneously (NIH's NCGC equipment), generally for pharmaceutical research.
- Protein interactions (Goodson et al., 2007) that involve simplifying process to allow for the parallel evaluation of protein interactions within a 96-well plate.

These systems address both of the core limitations of the traditional methods, in that they are scaled-up implementations that typically rely upon automation to conduct the assays. The obvious advantages of such systems are

1. Conducting a relatively large number of assays in parallel allows the researcher to work from the systems biology perspective because instead of starting with the lowest-level detail about a biological process—a single protein in this case—these methods enable large-view canvassing of a variety of proteins. The increased number of observations makes possible the detection of patterns that might otherwise have been lost.
2. Separating the practitioner from the physical experiment allows for more objectively repeatable experiments and results. Operator-specific technique is much less a variable in protein assays when the incubation, application, and washes are all controlled and administered by the machine itself.

The pervasive and thorny problem of how best to analyze such a significant quantity of protein assay data still remains, of course. Fortunately, many of the techniques described earlier within this text are designed specifically for high-volume data analysis, and much of the research work described in the following sections addresses specific issues of this data analysis problem directly.

## 14.4   INTEGRATION OF MOLECULAR DATA WITH HIGHER-LEVEL DATASETS

No man can learn what he has not preparation for learning, however near to his eyes is the object. A chemist may tell his most precious secrets to a carpenter, and he shall be never the wiser,—the secrets he would not utter to a chemist for an estate.

**Ralph Waldo Emerson**
*Spiritual Laws*

Specialization and generalization are the complementary extremes of the scientific process: It is only through the disciplined application of specialized knowledge and technique that the finest experiments can be conducted and the most detailed data generated, but it is only after the application of generalized, higher-level knowledge that the lower-level observations become meaningful. That is, without sufficient context, data are powerless. This is why the large life sciences data bases—be they dedicated to genes, proteins, diseases, what-have-yous—are maintained so diligently because they are shared repositories for the lowest-level data accumulated by scientists across the world that can only become useful once they are recognized and can be connected to a larger context. In systems biology, this often means relating data at a molecular level to higher-order process knowledge, whether it be a map of disease progression or gene activation networks or embryonic development; ideally, this also means making these connections assisted by software formalisms.

## 14.4.1 HIGHER-LEVEL DATA SETS AND PUBLISHERS

Higher-level data sets are often called *metadata*, in that they are structured information about lower-level data. Ontologies and taxonomies, for example, are artificial vocabularies with formal grammars that define precise memberships and relationships among terms. The groups that establish these formalisms are typically standards bodies, some of which are formally sanctioned and others of which are ad hoc collections of researchers sharing a common cause. Here are some representative examples of the higher-level data groups and the standards they promulgate:

- The National Center for Biomedical Ontology. They describe themselves as "a consortium of leading biologists, clinicians, informaticians, and ontologists who develop innovative technology and methods allowing scientists to create, disseminate, and manage biomedical information and knowledge in machine-processable form. Our vision is that all biomedical knowledge and data are disseminated on the Internet using principled ontologies, such that they are semantically interoperable and useful for improving biomedical science and clinical care."

- The Kanehisa Laboratories in the Bioinformatics Center of Kyoto University and the Human Genome Center of the University of Tokyo; the Kyoto Encyclopedia of Genes and Genomes (KEGG). KEGG is dedicated to "developing knowledge-based methods for uncovering higher-order systemic behaviors of the cell and the organism from genomic information."

- The systems biology markup language (SBML) is primarily for describing chemical reaction networks.

- The Microarray Gene Expression Data Society's Ontology Working Group. They describe their work as "charged with developing an ontology for describing samples used in microarray experiments."

- The HUPO Proteomics Standards Initiative, that "defines community standards for data representation in proteomics to facilitate data comparison, exchange, and verification."

- The Open Biomedical Ontologies Foundry. "The OBO Foundry is a collaborative experiment involving developers of science-based ontologies who are establishing a set of principles for ontology development with the goal of creating a suite of orthogonal interoperable reference ontologies in the biomedical domain."

- Gene Ontology Consortium; the Gene Ontology project. The project "is a collaborative effort to address the need for consistent descriptions of gene products in different databases. The project began as a collaboration between three model organism databases, FlyBase external link (*Drosophila*), the *Saccharomyces* Genome Database external link, and the Mouse Genome Database external link, in 1998. Since then, the GO Consortium has grown to include many databases, including several of the world's major repositories for plant, animal, and microbial genomes."

- National Center for Health Statistics; the International Classification of Diseases, Ninth Revision, Clinical Modification (ICD-9-CM). From their

Web site, "ICD-9-CM is the official system of assigning codes to diagnoses and procedures associated with hospital utilization in the United States. The ICD-9 is used to code and classify mortality data from death certificates."

### 14.4.2   TOOLS TO NAVIGATE METADATA

Some of the groups mentioned in the preceding section offer their own tools and Web sites to help traverse links among these data sets:

- KEGG has a site with multiple entry points to data, each of which has an extensive set of hyperlinks to other data sets; see http://www.genome.ad.jp/kegg/.
- The National Center for Biomedical Ontology offers a BioPortal, "a Web application to access the Open Biomedical Ontologies (OBO) library"; see http://www.bioontology.org/ncbo/faces/index.xhtml.
- The Microarray Gene Expression Data Society offers MAGE software and APIs; see http://mged.sourceforge.net/software/index.php.

Many of the higher-level data sets highlighted here are ontologies. Most of these take advantage of the work that the World Wide Web Consortium (W3C) has done to define standards around the *semantic web*, or formalized web of terms defined using formal vocabularies and grammars. The W3C says of the semantic web, it "is about two things. It is about common formats for integration and combination of data drawn from diverse sources, where the original Web is mainly concentrated on the interchange of documents. It is also about language for recording how the data relates to real world objects. That allows a person or a machine to start off in one database and then move through an unending set of databases which are connected not by wires but by being about the same thing."

There are tools dedicated just to working with these ontologies, including:

- Protégé, from Stanford University's Center for Biomedical Informatics Research, "is a free, open-source platform that provides a growing user community with a suite of tools to construct domain models and knowledge-based applications with ontologies. At its core, Protégé implements a rich set of knowledge-modeling structures and actions that support the creation, visualization, and manipulation of ontologies in various representation formats."
- Swoogle, from the UMBC ebiquity research group, is a semantic search engine that reports results from RDF/OWL documents found by crawlers over the Web; see http://swoogle.umbc.edu/.
- The Gene Ontology has a site that links to a number of different tools for working with ontologies; see http://www.geneontology.org/GO.tools.shtml.

### 14.4.3   ROLE OF HIGHER-LEVEL DATA IN SYSTEMS BIOLOGY

Consider the scientific method extended by connections to higher-level data sources, as depicted in Figure 14.2, below. Each of the core steps—hypothesizing, experimenting, and analyzing—has a tie to metadata stores.

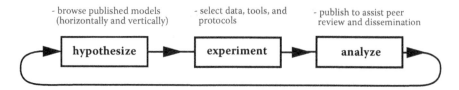

**FIGURE 14.2**   The scientific process empowered by higher-level data.

Hypothesizing or formulating the research question to address is aided by having the ability to review past work intelligently. The community has realized that—although reading research papers is extremely valuable—it is not necessarily the most natural way to sieve through large volumes of research data. Instead, standard groups have created formal mechanisms for organizing and interacting with data that allow both horizontal (reviewing lists of competing models, methods) and vertical (drilling into/out of models and methods to expose lower and higher levels of detail). Someone interested in studying prostate cancer, for example, might be interested in the problem from the point of view of SNP correlation, early detection, characterizing onset, the mechanics of metastasis, hormone regulation of the progressing disease, therapeutic interventions, responses to surgical events, etc. Groups such as the KEGG, the Microarray Gene Expression Data Society, the Open Biomedical Ontologies' OBO Foundry, and the HUPO Proteomics Standards Initiative all organize data to unite these different levels of investigation into a network of linked resources.

Planning and executing the experimentation benefits from having the ability to review the tools and methods used by other researchers, and to identify the best-of-breed solutions for the given research task.

Once the work is complete, ties to higher-level data sets also allow the practitioner to ensure that the data and outcomes are collected and published in a standard way. This allows the results to participate in the larger community, hopefully promoting increased visibility, validation (or challenge), and reuse.

Beyond the mechanics of planning and conducting research, there is also the bigger picture: Science is typically oriented toward answering *what* precisely happens to produce an *effect*; these effects are higher-order abstractions and require higher-order terminology.

## 14.5   IDENTIFYING GENES CONTROLLING MACROLEVEL CHANGES

Identifying genes that control macrolevel changes is one of the key goals of systems biology and is—albeit indirectly—the problem that Gregor Mendel was trying to solve when he experimented with peas, but although Mendel's macrolevel properties (phenotypes) were each motivated by a single gene, most of the phenotypes we wish to explain and control today are polygenic: controlled by more than one gene whose effects may be strongly influenced by environmental factors. Mendelian inheritance is not the help it might have been, so we rely on a different sort of search, one that is strongly influenced by the systems biology approach; briefly, given the target phenotype and the target population to study:

1. Generate a series of genetic markers to test among individuals
2. Identify markers that correlate with the target phenotype
3. Infer genes or gene expression patterns from the positively correlated markers

Each of these is addressed independently in the sections that follow.

### 14.5.1   GENERATE CANDIDATE MARKERS

It is useful to remember that among members of the same species, a large fraction of their genetic material is identical, suggesting that the differences we observe in phenotype are the result of contributions of a relatively minor percentage of their DNA (and environment). This is the core principle that motivates the International HapMap Project:

> Although any two unrelated people are the same at about 99.9% of their DNA sequences, the remaining 0.1% is important because it contains the genetic variants that influence how people differ in their risk of disease or their response to drugs. Discovering the DNA sequence variants that contribute to common disease risk offers one of the best opportunities for understanding the complex causes of disease in humans.

There are a number of common methods used to generate candidate markers from raw genetic material, including

- Amplified fragment length polymorphism (AFLP)
- Restriction-fragment length polymorphism (RFLP)
- Random amplified polymorphic DNA (RAPD)
- Microsatellites/simple sequence repeat (SSR)

These four methods are ways to reduce a single genotype into a much larger number of shorter fragments (markers). When these markers are all displayed together on a gel, the pattern of bands constitutes a DNA fingerprint of the individual, and images of these fingerprints can then be used to construct similarity-based hierarchies.

Note that, in species where markers are known to be largely homogeneous, suitable choices need to be made in the application of these methods to ensure that a larger number of markers are generated. This is more likely to be the case, for example, among commercially cultivated plants than it is for studies of human disease.

### 14.5.2   CORRELATE MARKERS WITH PHENOTYPE

As the fingerprinting methods in the preceding section suggest, the differences in markers among individuals will hopefully correlate with some portions of the pattern of incidence of the target phenotype. In the simplest case, there would be a clean partitioning: "Marker X is present if and only if the target condition is also present." This is almost never going to be the case, since many of the (commercially) interesting phenotypes are polygenic. So consider instead these three possible cases of interaction among influences that ought to be disentangled:

1. Additive: What possible combination of genes results in the minimum-acceptable threshold effect? Are the additive effects uniform, or do some markers correlate more strongly with an increase in the quantitative property than others?
2. Epistatic: The expression level of some genes may promote or inhibit or otherwise alter the effect of other genes. In this case, there is a temporal dependency among genes that needs to be identified. (This also influences the generation of markers, of course, because epistatis requires that the markers be generated based on expression data instead of nucleotide strings.)
3. Environmental: What influence does the environment have on the development of the target phenotype? If this is a crop study, can documentation about rainfall and fertilization help explain differences between samples better than differences between markers? Or is there a good way to correlate environmental differences with marker patterns?

For these correlations to be constructed reliably requires both good data and fairly rigorous analysis methods and software. It is also easy to see why botanical data are generally more receptive to analysis, both because detailed pedigrees are readily available and because the environmental data are maintained.

The result of this phase should be a list of markers, and for each marker, an identification of its responsibility in correlating with the quantitative trait: Is it additive? Does it have any appreciable influence on (or from) another marker? In what way is it affected by the environment? Similarity-based hierarchies (or other clustering maps) can be used to motivate hypothesized marker differences among similarity-presenting individuals.

### 14.5.3 INFER GENES OR GENE EXPRESSION PATTERNS

Because the markers initially were developed as relatively short segments of DNA randomly selected from the genome, the final phase of the effort is to try to identify where the significant markers belong. In essence, this means mapping markers to specific loci, allowing us to refer to them as quantitative trait loci (QTLs) instead of anonymous markers. One way to do this is to use linkage maps. A linkage map is constructed based on linkage disequilibrium, which says simply that the difference between the expected distribution of alleles at two locations and the observed distribution of alleles at these locations is primarily attributable to their proximity (see Lewontin and Kojima, 1960). This mathematical property allows for ordering of markers within chromosomes, provided that there is enough information about pedigree to inform the construction of the map. Validating the linkage map and the coverage of the genetic markers generated for the analysis is a length step in and of itself.

Once a reliable linkage map can be constructed, the goal becomes one of identifying which genes are closest to the effective QTLs. In cases where the full genotype is known, this might be as simple as inferring the gene directly from the locus information; in cases where the full genotype is not known, it probably means doing database searches on fragments around the key loci to identify similar genes in related species to infer what the active genes are at (and around) the target locations.

## 14.6 MOLECULAR-LEVEL IMAGE SYSTEMS AND SYSTEMS BIOLOGY

One of the challenges of systems biology is that it is necessarily more than the study of a large number of concurrent, interacting systems, but rather it is the study of a large number of molecules that may interact vastly faster and in more ways than we are able to conceive. It is not even possible to work out, with any fidelity, how one specific amino acid folds into a 3-D shape, let alone to identify where a specific instance of that protein might be created, how it might interact with the cytoplasm to fold, diffuse, become activated, denatured, and recycled. Multiply this problem by thousands upon thousands of concurrent species, each with multiple instances, and it is easy to see why the problem quickly becomes intractable. Yet, decomposition and isolation remain our best approaches to making sense of any process, even a collection of processes this convoluted. Advances in imaging and visual analytics help us to maintain the dichotomy of narrowing our field of view within the context of the entire collection of processes.

### 14.6.1 MOLECULAR-LEVEL IMAGING

There are a few properties of living organisms that make them particularly challenging to study from a systems perspective:

- Scale, space: From the micro to the macro, there are significant processes across spatial scales. Not only are many of the targets on the molecular scale, but their distribution in three-space also complicates both the collection and analysis of data. In addition, while there are imaging methods that can easily illustrate the macrolevel difference between bone and soft tissues, it is remarkably difficult at a smaller scale to differentiate between species of molecules.
- Scale, time: Alzheimer's disease may take years to develop, but proteins can change conformation thousands of times per second. Not only can it be challenging to capture data across these scales, but at the extreme, a huge volume of data are generated that subsequently must be catalogued and analyzed.
- In vivo: While many of the processes that we wish to study can be taken "off-line" as it were and studied in vitro, some of them can only be studied in active, live cells. This challenges many of the imaging modalities.

The lowest-level data, both in terms of physical scale and time scale, are observed using excitable tags and detection systems. Two choices for tags include fluorescent proteins and quantum dots. The former are relatively small, inert proteins with fluorescent motifs that can be attached at either terminus of a protein to be monitored and generally do not interfere with the conformation or the function of the tagged proteins. Quantum dots are crystalline semiconductors that are excited by UV light; they can be coated in biologically neutral agents to allow them to be injected into living tissues. These biological coatings can be primed so that they bind preferentially to specific proteins for tracking. The detection systems can include both direct imaging of fluorescence and the presence of induced fluorescence (in the case of FRET). These detectors can operate at a speed that makes tracking molecular migrations feasible. The disadvantages of these methods are the cost of the equipment involved, the application of the tags, and the processing of the data.

At a higher-level, laser-based and tomographic methods can be used to study intercellular and intracellular structures, chemical pathways (by tracking macromolecules), and membrane potentials and gradients. The disadvantages of these methods are (again) the cost of the equipment, the increased complexity of using the methods (especially in the case of tomographic and FFT-based designs), and processing the large volume of data.

There continues to be a large and active research community around biological imaging, however, so both the cost and the complexity of generating rich, in vivo, three-dimensional images over time ought to decrease significantly, leaving us with the sole difficulty of figuring out what to do with such a glut of data.

### 14.6.2 Visual Analytics

Visualization once was the goal of many software applications, presumably because the use was considered sufficiently well equipped to understand precisely what was being displayed on the screen. Visual analytics as a discipline assumes otherwise: The user is, in fact, remarkably well equipped to interpret data, but some of our innate facility to understand data is tied directly to our ability to interact with data. In much the same way that we can see a complicated series of gears and reason through what happens when the handle is turned, we do not necessarily believe it—and often prefer simply to take the shortcut—of reaching out and turning the handle to see what *really* happens. Visual analytics provides us with this capability: Display data in a clever way but allow us the freedom to interact with the visualization so that we can sift, sort, filter, pivot, and dilate the information in whatever ways it takes to refine our understanding of what we are seeing.

Visual analytics is proving to be very successful in the intelligence community. The U.S. Department of Homeland Security established the National Visualization and Analytics Center to pursue research in this direction. Fortunately, many of these same analysis tasks are similar to the systems biology practitioner:

- Highly networked data: Instead of networks of terrorist cells, biologists work with networks of living cells and highly networked protein-protein interactions.
- Highly ordered temporal data: Instead of sequences of wire-fund transfers, biologists work with sequences of gene activation and repression and cell membrane channel openings and closings.
- Massive amounts of data: The number of cell phone calls placed in a year pales in comparison to the number of times and ways a single calcium ion is used within a biological network in a day.

- Probabilistic data: In much the same way that the word "fear" will be used in positive and negative examples of clandestine messages, tumor necrosis factor appears in both healthy and unhealthy cellular mechanisms; a thorough exploration of the data is required to disambiguate.

Visual analytics is proving to be a very powerful tool for bioinformaticists to explore data sets, particularly those arising from low- or molecular-level data, and generate hypotheses.

## 14.7   SUMMARY

Systems biology is a sense-making approach to understanding a very complex, adaptive system. It is also an area for very active and far-reaching research as we have seen throughout this treatment. The deeper one looks—whether that is deep in terms of reducing the scale of investigation to the lifespan of individual molecules or deep in terms of the increasing scope of the process map we are trying to build—the core principals of a systems biology approach still apply: Large, interrelated problems require large volumes of data and significant manual and automatic analysis methods to produce useful results. Sometimes this means modifying procedures to use microarray data to examine the contents of a single cell; at other times, it means using a roboticized, streaming device to process large numbers of assays in parallel; often it means traversing through different scales, starting with molecular data, and tying it back to macroscopic process paths or genetic products; it may even mean relying on visual analytic solutions to interact with the data, so that sense-making is neither entirely in the machine nor entirely outside but occurs at the interface between the two.

Hopefully, it is at precisely this interstice where the systems biology approach becomes most powerful.

## 14.8   PROBLEMS

1. Contrast RT-PCR and quantitative RT-PCR. How do these differ from T7 amplification?
2. What is ELISA, and what is it used for?
3. What is SBML? Under what circumstances might you either encounter it or use it? Are these standards for machine use, personal use, or somewhere in between?
4. What is AFLP, and why is it popular? How is it different from
   a.   RFLP?
   b.   RAPD?
   c.   SSR?
5. What is QTL mapping, and how is it accomplished?
6. What is a cM, and how does this relate to linkage maps and linkage disequilibrium? LOD?
7. What is OMI, and what is its relationship to systems biology?

## REFERENCES AND NOTES

Bowen, R. A. "Biotechnology and Genetic Engineering." Available at http://www.vivo.colo state.edu/hbooks/genetics/biotech/index.html. January 3, 2000.

Caliper LifeSciences. Available at http://www.caliperls.com/products/labchip-systems/lab chip90.htm.

Campbell, A. Malcolm, "ELISA (Enzyme-Linked ImmunoSorbant Assay)." Available at http://www.bio.davidson.edu/Courses/genomics/method/ELISA.html.

Clontech, "Overview: cDNA Synthesis Using SMART™ Technology." Available at http://www.clontech.com/products/detail.asp?product_family_id=1415&product_group_id=1430& product_id=169243.

Ding, Li, Tim Finin, Anupam Joshi, Rong Pan, R. Scott Cost, Yun Peng, Pavan Reddivari, Vishal C Doshi, and Joel Sachs, "Swoogle: A Search and Metadata Engine for the Semantic Web." In *Proceedings of the Thirteenth ACM Conference on Information and Knowledge Management*, November 9, 2004.

Dove, Alan, "High-Throughput Screening Goes to School." *Nature Methods,* 4 (2007): 523–532, Available at http://www.nature.com/nmeth/journal/v4/n6/full/nmeth0607-523.html.

Du, Wei, Ying Wang, Qingming Luo, and Bi-Feng Liu, "Optical Molecular Imaging for Systems Biology: From Molecule to Organism." *Analytical and Bioanalytical Chemistry* (2006): 444–457.

Engvall, Eva, and Peter Perlman, "Enzyme-Linked Immunosorbent Assay (ELISA) Quantitative Assay of Immunoglobulin G." *Immunochemistry,* 8 (1971): 9, 871–874.

Esumi, Shigeyuki, Sheng-Xi Wu, Yuchio Yanagawa, Kunihiko Obata, Yukihiko Sugimoto, and Nobuaki Tamamaki, "Method for Single-Cell Microarray Analysis and Application to Gene-Expression Profiling of GABAergic Neuron Progenitors." *Neuroscience Research,* 60 (2008): 4, 439–451.

Ewens, W. J., and R. S. Spielman, "The Transmission/Disequilibrium Test: History, Subdivision, and Admixture." *Human Genetics,* 57 (1995): 2, 455–464.

The Gene Ontology, "An Introduction to the Gene Ontology." Available at http://www.geneon tology.org/GO.doc.shtml.

Gold, David L., Kevin R. Coombes, Jing Wang, and Bani Mallick, "Testing Gene Class Enrichment in High-Throughput Genomics." *Brief Bioinformatics* 8 (2007): 2, 71–77.

Goodson, Michael L., Behnom Farboud, and Martin L. Privalsky, "An Improved High Throughput Protein-Protein Interaction Assay for Nuclear Hormone Receptors." *Nuclear Receptors Signaling,* 5 (2007): e002. Available at http://www.pubmedcentral.nih.gov/articlerender.fcgi?artid=1853068.

Hermjakob, Henning, Luisa Montecchi-Palazzi, Gary Bader, Jérôme Wojcik, Lukasz Salwinski, Arnaud Ceol, Susan Moore, Sandra Orchard, Ugis Sarkans, Christian von Mering, Bernd Roechert, Sylvain Poux, Eva Jung, Henning Mersch, Paul Kersey, Michael Lappe, Yixue Li1, Rong Zeng, Debashis Rana, Macha Nikolski, Holger Husi, Christine Brun, K. Shanker, Seth G. N. Grant, Chris Sander, Peer Bork, Weimin Zhu, Akhilesh Pandey, Alvis Brazma, Bernard Jacq, Marc Vidal, David Sherman, Pierre Legrain, Gianni Cesareni, Ioannis Xenarios, David Eisenberg, Boris Steipe, Chris Hogue, and Rolf Apweiler, "The HUPO PSI's Molecular Interaction Format—A Community Standard for the Representation of Protein Interaction Data." *Nature Biotechnology,* 22 (2004): 177–183.

High-Throughput Screening. Available at http://en.wikipedia.org/wiki/High-throughput_screening.

HTG, "High-Throughput Genomics, Inc." Available at http://www.htgenomics.com/.

Hucka, M., A. Finney, H. M. Sauro, H. Bolouri, J. C. Doyle, H. Kitano, and the SBML Forum: A. P. Arkin, B. J. Bornstein, D. Bray, A. Cornish-Bowden, A. A. Cuellar, S. Dronov, E. D. Gilles, M. Ginkel, V. Gor, I. I. Goryanin, W. J. Hedley, T. C.

Hodgman, J.-H. Hofmeyr, P. J. Hunter, N. S. Juty, J. L. Kasberger, A. Kremling, U. Kummer, N. Le Novère, L. M. Loew, D. Lucio, P. Mendes, E. Minch, E. D. Mjolsness, Y. Nakayama, M. R. Nelson, P. F. Nielsen, T. Sakurada, J. C. Schaff, B. E. Shapiro, T. S. Shimizu, H. D. Spence, J. Stelling, K. Takahashi, M. Tomita, J. Wagner, and J. Wang. "The Systems Biology Markup Language (SBML): A Medium for Representation and Exchange of Biochemical Network Models." *Bioinformatics*, 19 (2003): 524–531.

The International HapMap Project. Available at http://www.hapmap.org/index.html.en.

Iscove, N. N., M. Barbara, M. Gu, M. Gibson, C. Modi, N. Winegarden, "Representation Is Faithfully Preserved in Global cDNA Amplified Exponentially from Sub-picogram Quantities of mRNA." *Nature Biotechnology*, 20 (2002): 9, 940–943.

"Meeting Report—From High Throughput Genomics to Useful Transgenic Crops." *Molecular Breeding*, 5 (October 1999): 481–483.

Johnson, V. P. and Carol Christianson, "Multifactorial Inheritance." In *Clinical Genetics: A Self-Study Guide for Health Care Providers*. Available at http://www.usd.edu/med/som/genetics/curriculum/1GMULTI5.htm.

KEGG: Kyoto Encyclopedia of Genes and Genomes. Available at http://www.genome.ad.jp/kegg/.

Lewontin, R. C., "The Interaction of Selection and Linkage. I. General Considerations: Heterotic Models." *Genetics*, 49 (1964): 49–67.

Lewontin, R. C., "On Measures of Gametic Disequilibrium." *Genetics*, 120 (1988): 849–852.

Lewontin, R. C., and K. Kojima, "The Evolutionary Dynamics of Complex Polymorphisms." *Evolution*, 14 (1960): 458–472.

Martens, L., S. Orchard, R. Apweiler, and H. Hermjakob, "Human Proteome Organization Proteomics Standards Initiative: Data Standardization, a View on Developments and Policy. *Molecular and Cellular Proteomics*, 6 (2007): 1666–1667.

Meudt, Heidi M., and Andrew C. Clarke. "Almost Forgotten or Latest Practice? AFLP Applications, Analyses and Advances." *TRENDS in Plant Science,* 12 (2007): 3, 106–107.

Microarray Gene Expression Data Society (MGED) Network, "Ontology Working Group." Available at http://mged.sourceforge.net/ontologies/OntologyResources.php.

Mueller, Ulrich G., and L. LaReesa Wolfenbarger, "AFLP Genotyping and Fingerprinting." *Trends in Ecology and Evolution*, 14 (1999): 10, 389–394.

Nagorsen, Dirk, Ena Wang, and Monica C. Panelli, "High-Throughput Genomics: An Emerging Tool in Biology and Immunology." *ASHI Quarterly* (Third Quarter 2003). Available at http://www.ashi-hla.org/publicationfiles/ASHI_Quarterly/27_3_2003/3rd_qtr_Genomics.pdf.

The National Center for Biomedical Ontology. Available at http://www.bioontology.org/.

National Center for Health Statistics, "International Classification of Diseases, Ninth Revision, Clinical Modification (ICD-9-CM)." Available at http://www.cdc.gov/nchs/about/other act/icd9/abticd9.htm.

National Institutes of Health (NIH) Chemical Genomics Center (NCGC), "Assay Guidance// Assay Guidance Manual//Enzymatic Assays." ©2008, Eli Lilly and Company and the National Institutes of Health Chemical Genomics Center. Available at http://www.ncgc.nih.gov/.

National Visualization and Analytics Center. Available at http://nvac.pnl.gov/.

The Open Biomedical Ontologies. Available at http://www.obofoundry.org/.

Orchard, S., L. Montecchi-Palazzi, H. Hermjakob, and R. Apweiler, "The Use of Common Ontologies and Controlled Vocabularies to Enable Data Exchange and Deposition for Complex Proteomic Experiments." *Pacific Symposium on Biocomputing* (2005): 186–196. http://www.scientistlive.com/lab/?/Equipment/2005/12/21/14659/Platform_for_high-throughput_automation_of_protein_sizing/.

Razvi, Enal S., and Lev J. Leytes, "High-Throughput Genomics: How SNP Genotyping Will Affect the Pharmaceutical Industry." *Modern Drug Discovery*, 3 (2000): 5, 40–42.

Sastry, Srinivas S., and Barbara M. Ross, "Nuclease Activity of T7 RNA Polymerase and the Heterogeneity of Transcription Elongation Complexes." *Journal of Biological Chemistry*, 272 (1997): 8644–8652.

Stanford Center for Biomedical Informatics Research (protégé-owl). Available at http://pro tege.stanford.edu/overview/protege-owl.html.

Systems Biology at Pacific Northwest National Laboratory (PNNL), "Advanced Cell and Molecular Imaging." Available at http://www.sysbio.org/capabilities/cellimaging/.

Tajiri, Kazuto, Hiroyuki Kishi, Yoshiharu Tokimitsu, Sachiko Kondo, Tatsuhiko Ozawa, Koshi Kinoshita, Aishun Jin, Shinichi Kadowaki, Toshiro Sugiyama, and Atsushi Muraguchi, "Cell-Microarray Analysis of Antigen-Specific B-Cells: Single Cell Analysis of Antigen Receptor Expression and Specificity." *Cytometry Part A*, 71A (2007): 11, 961–967.

Tiwari, Abhishek, and Arvind K. T. Sekhar, "Workflow Based Framework for Life Science Informatics." *Computational Biology and Chemistry*, 31 (2007): 305–319.

Van Berloo, Ralph, Aiguo Zhu, Remco Ursem, Henk Verbakel, Gerrit Gort, and Fred Antonis van Eeuwijk, "Diversity and Linkage Disequilibrium Analysis Within a Selected Set of Cultivated Tomatoes." *Theoretical and Applied Genetics*, 117 (2008): 89–101.

Van Ooijen, J. W., "JoinMap® 4, Software for the Calculation of Genetic Linkage Maps in Experimental Populations." Wageningen, Netherlands: Kyazma B.V. (2006). Available at http://www.kyazma.nl/index.php/mc.JoinMap/sc.Evaluate/.

Yamamura, Shohei, Hiroyuki Kishi, Yoshiharu Tokimitsu, Sachiko Kondo, Ritsu Honda, Sathuluri Ramachandra Rao, Masahiro Omori, Eiichi Tamiya, and Atsushi Muraguchi, "Single-Cell Microarray for Analyzing Cellular Response." *Analytical Chemistry*, 77 (2005): 24, 8050–8056.

Zhang, Xuegong, "Machine Learning in High-Throughput Genomics and Proteomics." *13th International Conference on Neural Information Processing, October 3–6, 2006, Hong Kong*. Available at http://iconip2006.cse.cuhk.edu.hk/program/Notes/zhang1.pdf.

# Index

21/10/2024

01777107-0001